THE GREAT LIVES SERIES

Great Lives biographies shed an exciting new light on the many dynamic men and women whose actions, visions, and dedication to an ideal have influenced the course of history. Their ambitions, dreams, successes and failures, the controversies they faced and the obstacles they overcame are the true stories behind these distinguished world leaders, explorers, and great Americans.

Other biographies in the Great Lives Series

CHRISTOPHER COLUMBUS: The Intrepid Mariner

MIKHAIL GORBACHEV: The Soviet Innovator

JOHN F. KENNEDY: Courage in Crisis

ABRAHAM LINCOLN: The Freedom President

HARRIET TUBMAN: Call to Freedom

ACKNOWLEDGMENT

A special thanks to educators Dr. Frank Moretti, Ph.D., Associate Headmaster of the Dalton School in New York City; Dr. Paul Mattingly, Ph.D., Professor of History at New York University; and Barbara Smith, M. S. , Assistant Superintendent of the Los Angeles Unified School District, for their contributions to the Great Lives Series.

SALLY RIDE
SHOOTING FOR THE STARS

Jane Hurwitz and Sue Hurwitz

FAWCETT BOOKS
NEW YORK

For middle school readers

A Fawcett Book
Published by The Random House Publishing Group

Published in the United States by Fawcett Books, an imprint of The Random House Publishing Group, a division of Random House, Inc., New York and simultaneously in Canada by Random House of Canada Limited, Toronto.

FAWCETT BOOKS and colophon are trademarks of Random House, Inc.

Library of Congress Catalogue Card Number: 89-90821

ISBN: 0-449-90394-X

Cover design and illustration by Paul Davis

Manufactured in the United States of America

www.ballantinebooks.com

First Edition: September 1989

19 18 17 16 15

TABLE OF CONTENTS

1

Lift-Off into Space

KENNEDY SPACE CENTER: June 18, 1983. The astronaut van stopped in front of the launchpad and Dr. Sally Ride, mission specialist, and the four other members of the crew peeked out the window. A gentle Atlantic breeze blew across her face, stirring her curly brown hair, but Sally didn't notice. Spotlights bounced off the gleaming white exterior of the space shuttle, *Challenger*. Sally's attention was riveted to the awesome sight of the magnificent flying machine shrouded in Florida's predawn darkness. This was the greatest moment in her life!

The *Challenger* stood on its tail with its nose pointed toward the sky, the very same sky that Sally and the other four crew members would be thundering through in several hours. As Sally stepped down from the van and walked closer to the *Challenger*, the sputtering and gurgling of the shuttle's engines echoed the bubbling of her own thoughts. She was finally getting her chance to fly in space!

As Sally contemplated her first spaceflight, she wondered how she would react. As part of her extensive training for the shuttle flight, she had already spent countless hours in a simulator of the shuttle cockpit where she had experienced small doses of weightlessness. Still, she wondered how weightlessness, experienced while actually traveling in space, would really feel. How would her body react to the force of gravity pulling against her during lift-off? How would she react to the confinement of the small shuttle quarters during six days in space?

As technicians hovered around the orbiter, carefully reexamining its many complicated components, Sally's thoughts raced through the events in her life that led up to this moment. Long years studying math and science had trained her mind in problem solving. Intensive sports competition had trained her body. Now she had to prove she could put all that experience to good use.

The scary truth was that Sally was about to be strapped into a four-million-pound aluminum and ceramic tile flying machine that stood over thirty stories high and was powered by a dangerous propellant. Although many people believed that space shuttle flights were nearly as safe as flying in commercial jets, Sally knew otherwise.

In March 1981, when a test on the space shuttle *Columbia* was being conducted on one of the launchpads, an accident occurred. Unexpectedly, a compartment inside the rear of the shuttle was enveloped in gaseous nitrogen, depleting most of its oxygen. Even though the environment inside the chamber was clearly hazardous, an "all clear" signaled that it was safe for five waiting technicians to enter. When they did, disaster struck — the workers collapsed from the poisonous fumes. The five were rushed to the hospital, but two died from inhaling the noxious gas. Sally knew that there were risks to spaceflight, but she was also thrilled about being propelled two hundred miles above Earth. Like the other astronauts with her, Sally believed that the dangers were worth the reward of doing something new — breaking new ground, exploring the unknown. Sally would also be the first American woman to fly in space, though this was not her main focus.

The pungent smell of jet fuel mingled with the salty air of coastal Florida as Sally and her companions walked to the elevator and stepped inside for the ride up to the shuttle's cabin. The slow, tension-filled ascent up the scaffold beside the 195-foot shuttle was shared by Commander Robert Crippen, Pilot Frederick Hauck, Medical Doctor Norman Thagard (the first medical doctor assigned to a shuttle mission), and the other mission specialist, John Fabian. Though this was the seventh shuttle mission, there would be many new experiments, procedures, and tasks to carry out once the crew was in space.

As they left the elevator the astronauts were greeted by several technicians, who scurried around checking and rechecking the astronauts' flight suits — especially their air supply — and making sure they were staying on schedule. Sally's anticipation mounted as the technicians helped her slip on her launch helmet, which would provide her with oxygen and radio communication with Mission Control during lift-off.

After Sally and the other crew members climbed through the small hatch of the *Challenger*'s nose cone into the flight cabin, technicians prepared to strap the astronauts into their reclining seats. Sally's eyes scanned across the flight deck of the shuttle orbiter. It looked much like that of a large commercial airline, except that three video screens occupied the center of the instrument console. These screens displayed vital information about the shuttle's vast and complicated flight and navigation systems. Sally had practiced using the shuttle's instruments many times; her training was ready to be put to the test.

Once Sally was lying back in her seat facing the nose of the shuttle, the awkward perspective made it difficult to look out the windows. Still, the position would help her cope with the tremendous acceleration of lift-off and ascent into orbit that the shuttle would go through in order to blast its way out of Earth's gravitational pull.

While the actual launch was still some time away, Sally could feel her heart pounding and her blood rushing to her head.

Sally knew that her parents, sister, and husband were watching in a special room provided for the astronauts' families and others with a personal interest in the launch. Even though she could not see her loved ones — including Steve Hawley, her husband and a fellow astronaut — Sally knew that their thoughts and prayers were focused on her and the historic flight. As dawn broke and rays of sunlight crept over the horizon, more than 250,000 spectators crowded the beaches and causeways around the Kennedy Space Center in Florida. Some spectators wore T-shirts with the slogan, Ride, Sally Ride. Clenched fists, gritted teeth, and other signs of nervousness were visible along with joyous cheers as the onlookers waited impatiently for the moment of lift- off.

Inside the shuttle, the voice of the ground controller from Mission Control played counterpoint to the various mechanical noises of the warming engines. As every valve and throttle was checked and rechecked by computers, the astronauts watched the many lights on the control panel blink on and off like elaborate Christmas decorations. For Sally, this ride into space was better than any Christmas present.

As a child Sally had always been fascinated by the planets, stars, and galaxies, but she had never dreamed of becoming an astronaut. Her interest in science had led instead to a degree in astrophysics, through which she had learned about the characteristics and properties of stars. Now, at the age of thirty-two, she was about to become the youngest American astronaut ever to go on a space mission.

At 7:33 A.M. the countdown to launch entered its last few critical seconds. Sally heard the *Challenger*'s mighty engines roar to life, as the tremendous rockets flanking the huge fuel tank of the shuttle ignited and the orbiter began to quiver. As a gigantic cloud of dark smoke spewed from *Challenger*'s solid rocket boosters, Sally breathed deeply — her six-day, two-hour voyage into the heavens had begun.

The crowd of spectators cheered the astronauts off on their voyage. Sally was busy monitoring the flight controls. The on-board computer lights were blinking madly and flight information was being processed at an incredible speed. Sally needed every ounce of concentration she had to keep track of the changes and call out data to the other astronauts.

The powerful rocket engines built up to the thrust of 450,000 pounds needed to lift the *Challenger* and its crew beyond the pull of the Earth's gravity and into orbit. The pressure, or G-force, was so great it was as if the planet did not want the ship or astronauts to leave. Sally's head vibrated from side to side inside her helmet and the noise from the rockets blotted out nearly all other sound. Indeed, she had to strain her neck to lift

A picture-perfect lift-off of the space shuttle Challenger. *The six-day mission in 1983 featured two satellite launchings, various experiments using the "remote manipulator system," and Dr. Sally Ride, the first American woman in space.*

her head in order to read the data on her bank of instruments.

Two minutes later, after the booster rockets burned out, the ride became steady and calm, even though the spacecraft was now traveling at a speed of about 17,400 miles an hour — or nearly 5 miles per second! Two smaller engines ignited at the rear of the orbiter, thrusting the *Challenger* still higher and higher. Forty-four minutes and 23.7 seconds after lift-off, the final orbiting position was reached — about 200 miles above the Earth, where the astronauts' real job would begin. Over the next six days, Sally and the rest of the crew would collect vast amounts of scientific data, use a robot arm to deploy and retrieve satellites, and conduct a variety of experiments.

Dr. Sally Ride's participation in the flight of the space shuttle *Challenger* was not only a personal achievement, but one which would help to open up new opportunities for other American women in a variety of professions. Her courage, determination, and commitment to teamwork earned her respect from her fellow astronauts as well as from millions of Americans — and a place in the history of space exploration.

2

A Childhood of
Tennis and Science

SALLY KRISTEN RIDE was born in Los Angeles, California on May 26, 1951, to parents who valued education. Sally's father, Dr. Dale Ride, was a political science professor at Santa Monica College and later became assistant to the president and superintendent there. Sally's mother, Joyce Ride, was a teacher, voracious reader, and self-motivated intellectual.

Sally and her younger sister, Karen, were raised in an atmosphere that encouraged individual exploration. Accordingly, Sally believed that she could undertake any activity that she felt capable of or wished to learn about. Being a girl never prevented her from doing anything she wanted. While Sally's parents had no particular talent in science or athletics, their open attitude toward learning and personal achievement allowed Sally's intense love of sports and science to grow unrestricted. From the beginning of her life, then, Sally Ride aimed to shoot for the stars.

When asked what they had done to foster achievement in their daughters, Sally's father once responded, "We just let them develop normally. We might have encouraged, but mostly we let them explore."

Sally's mother added, "In a way you could look at it as neglect. Dale and I simply forgot to tell them that there were things they couldn't do. But I think if it had occurred to us to tell them, we would have refrained. . . . We didn't talk a lot about careers because we didn't want to pressure them into anything specific."

Free to explore, and permitted to be herself, Sally benefited from her comfortable, cultured home life. She grew up reading Nancy Drew mysteries, James Bond spy novels, and a good deal of science fiction. Superman was among her favorite heroes, and she could recite the entire prologue from the old television show: "Superman! Faster than a speeding bullet. . . more powerful than a locomotive. . . able to leap tall buildings at a single bound!" Sally also loved sports. Her father remembers that Sally read the sports pages from the time she was five years old and had an impeccable memory for baseball statistics.

Indeed, Sally excelled in sports and was so good at playing softball and football that often she was the only girl selected to participate in traditionally all-boy street games — and was often the top scorer in the games she played. Sally learned early two significant lessons: the importance of being a team player, and that girls can compete in boy's games. It didn't matter to Sally that some people thought it was unusual to see a girl playing on a boy's team. She saw only that there was a game to play, that she aimed to be the best she could at playing it, and that she wanted to win.

Her mother pointed out, "The main thing was that if Sally was interested in a subject, she'd give it all her attention. If she wasn't interested, she didn't give it her attention. She sets her own goals and competes with herself."

The years of Sally's childhood coincided with great changes in the field of space exploration. On October 4, 1957, the Soviet Union sent *Sputnik I,* the first artificial satellite, into orbit around the earth, taking an impressive step toward their publicly stated goal of exploring this new frontier. The world was shocked, and Americans in particular were stunned that the Soviets had taken the lead in what marked the beginning of the Space Age. Many people said that the country needed to make a true commitment to space exploration and to the development of the necessary technology. As a result, in 1958 the National Aeronautics and Space Administration (NASA) was established as an independent agency to conduct space experiments and exploration. A quarter century later, Sally Ride would fly in the space shuttle for that same agency.

In 1960, when Sally was nine years old, her father took a leave of absence from his work so that the family could spend a year traveling in Europe. While they traveled, Sally and her sister were tutored by their parents so that they wouldn't get behind in their studies. Indeed, when they returned, Sally moved at a pace that was quicker than the rest of her class. The Rides bought a home in Encino, near Los Angeles, and Sally continued to expand both her interest in science and her devotion to sports.

In 1961 man's age-old dream of traveling beyond Earth's limits took another giant leap forward. A Soviet cosmonaut, Yuri Gagarin, became the first person to travel in space, orbiting the earth once in a flight that lasted 1 hour and 48 minutes. President John F. Kennedy, challenged by this new sign of Soviet prowess, pledged that the United States would land a man on the moon by the end of the decade. His competitive spirit captured the imagination of many people, among them Sally Ride.

Though she studied hard in school, and particularly enjoyed learning about the stars, planets, and solar system, Sally's first love was sports. She even fantasized about joining the Los Angeles Dodgers baseball team, but her parents suggested that she take up tennis instead. Little did they realize that this would end up preparing her for future success as an astronaut.

12

Sally began taking tennis lessons from Alice Marble, a four-time women's national champion. She practiced hard, set high goals for herself, and soon was winning many weekend and summer tournaments. In fact, her determination and talent eventually made her the eighteenth-ranked junior player in the United States. It was this same perseverance and high motivation that she would use later during her training as an astronaut.

In February 1962 Sally and millions of other Americans cheered astronaut John Glenn when he became the first American to orbit the Earth. His three revolutions took 4 hours, 55 minutes and gave a tremendous boost to the newly aggressive U.S. space program. The United States was gaining on the Soviets in the race for the moon and, as the country's interest in space increased, so did Sally's.

Sally spent much of her time at Portola Junior High School studying science and math. One of Sally's teachers, Wilbur Hansen, remembers her well. "In my class, Sally was quiet, did her work, and was very private," he recalls. "I wouldn't say she was remarkable in any way — she was just a good student."

In 1964 Sally won a partial tenth-grade scholarship to Westlake, an exclusive private girls' school in Beverly Hills, California. It was there that she met Dr. Elizabeth Mommaerts, a science teacher who introduced her to the scientific method of solving problems through the skills of

13

logical deduction and reasoning. Sally so admired Dr. Mommaerts's approach to learning that she tried to model her own actions after her. Dr. Mommaerts, for her part, was so impressed by Sally's scholastic abilities that she encouraged her to become a scientist. Although Sally spent many hours honing her tennis skills, she found the time to broaden her high school studies to include chemistry, physics, trigonometry, and calculus — courses which gave her a solid background for her college studies and, later, her training as an astronaut.

In 1968 Sally Ride, like many other Americans, read newspapers and watched television reports describing the frenetic race for the moon between the Soviets and the United States. In December *Apollo 8* became the first manned spacecraft to reach the moon; after ten orbits the ship returned home triumphant after a six-day, three-hour journey.

Sally, now seventeen, was one of the top six students in her graduating class at Westlake School, even though she was a year younger than her classmates. The following September she left southern California to attend Swarthmore College, a small liberal arts university just outside of Philadelphia, Pennsylvania. Although she entered the college as a physics major, she quickly made a name for herself as an excellent athlete by winning back-to-back crowns in the Eastern Intercollegiate Woman's Tennis Championships in her freshman and sophomore years. In fact, though she devoted many hours to her science and literature studies, tennis was still such a passion that she seriously considered becoming a professional. A further sign of her seriousness in this direction came when she decided, because Swarthmore had no indoor tennis courts — preventing her from playing tennis year-round — to leave the school after three semesters to return to sunny California.

That year, 1969, was a turning point for the United States' space program. On July 20, as millions of people around the world watched on television, two American astronauts from the spacecraft *Apollo 11* became the first humans to land on and explore the moon's surface. They spent two hours "moon walking," setting up instruments to leave on the lunar surface, and collecting soil and rock samples before reentering their lunar module, returning to their command module, and making the journey of about 240,000 miles back to earth. The nine-day mission left an indelible mark on the future of space exploration, on humankind, and on Sally Ride. She shared the pride and wonder that most people felt toward this extraordinary scientific achievement. At one point, after she had become an astronaut, Sally said, "I've always watched the space program closely. I could tell you exactly where I was when John Glenn went into space and when Neil Armstrong walked on the moon."

Still, though Sally's interest in space had grown considerably in the last few years, and she was taking a physics course at the University of California at Los Angeles, she was not considering a career as an astronaut, nor even one as a scientist. Tennis remained her number-one passion. During this time, Sally met Billie Jean King, a Wimbledon champion and one of America's greatest women tennis stars. Billie Jean thought so highly of Sally's potential as a tennis professional that she advised her to pursue the sport as a career. After several more months of intensive training, however, Sally reluctantly came to the conclusion that she did not have what it takes to become a tennis professional and decided to give up the idea. Sally's mother explained her daughter's decision this way: "Sally simply couldn't make the ball go just where she wanted it to. And Sally wouldn't settle for anything short of excellence in herself."

Sally refocused her efforts on becoming a scientist. She returned to college, choosing Stanford University, near San Francisco, California. There she studied science and literature — particularly Shakespeare. "I really had fun reading Shakespeare's plays and writing papers on them," Sally commented later. "It's kind of like doing puzzles. You had to figure out what he was trying to say and find all the little clues inside the play that [proved] you were right."

Between 1969 and 1973 Sally followed the NASA space program with increased interest. In 1972 President Richard Nixon authorized the development of the space shuttle program, a

manned transportation system that would re-
duce the cost of using space for commercial, scientific, and military purposes. There were five
more lunar landings and the launching of Skylab,
America's first orbiting space station, on May 25,
1973. Project Skylab was of particular interest to
would-be astronauts because it was designed to
demonstrate that people could live and work in
space for prolonged periods of time without suf-
fering ill effects. Over the duration of the program, American astronauts spent as many as
eighty-four days in space conducting experiments. The success of Skylab helped facilitate
the development of the space shuttle program.

In June 1973 Sally graduated from Stanford
University with two degrees: a bachelor of arts
degree in English and a bachelor of science degree in physics. In September she entered
Stanford's master's of science program in physics, specializing in astrophysics, the study of the
physical and chemical characteristics of celestial
matter. Following her childhood fascination with
astronomy, Sally focused on studying the X rays
given off by stars. Two years later Sally received
her master's degree, but she was still not finished
with her education. Before she could be considered a scientist — an astrophysicist — Sally had
to earn a doctorate degree, or Ph.D., which meant
conducting some original research, something
that no one had ever done before.

Speaking to a group of youngsters about how
she became an astronaut, Sally reminisced about

what she thought she wanted to do upon graduation: "I was interested in space but it wasn't anything I built a career around. Instead I planned to go into research in physics. I wouldn't have known how to prepare for a career as an astronaut even if it had occurred to me to try, since women weren't involved in the space program at that time."

3

Help Wanted: Astronauts and Mission Specialists

IN 1977 SALLY Ride was at a crossroads. Like most graduate students who reach the end of a long college career, Sally began to look for a job. Nearly finished with her doctorate in astrophysics, Sally hoped that she could find employment in which her highly specialized training could be put to good use.

One day, while looking in the help wanted section of the *Stanford Daily*, the university's newspaper, Sally came across a job advertisement that would change her life. NASA was looking for young scientists to become "mission specialists" who would conduct experiments on board space shuttle flights. The space agency said they wanted people who were flexible enough to pursue excellence outside their chosen field — and

NASA was urging women to apply! This was a distinct change in their policy, for historically NASA had accepted only military test pilots, and later scientists — all men — as astronaut candidates.

This change reflected the changing climate in the United States regarding women's rights. Women's rights had become an important and popular issue, covered frequently and in-depth by the national media. A national women's political conference held in Houston, Texas to discuss the proposed Equal Rights Amendment was attended by three First Ladies: Rosalynn Carter, Betty Ford, and Lady Bird Johnson. Inspired by the attention given to equality for women, especially in the workplace, Sally began to believe that it wouldn't be long before women were accepted into NASA's astronaut program, and she was right.

Carolyn L. Huntoon, the Johnson Space Center's deputy chief of personnel and a member of the astronaut selection committee, put it this way: "Because we had a new spacecraft, and it was going to be built so that it had space inside it . . . and could have toilet facilities that could accommodate women . . . and, I think, because at that time in our country, people were feeling a little bit bad about the way they treated women . . . they said, 'It's a federal job and we're going to open it to all races, sexes, religious backgrounds and ages.' "

NASA was also at a crossroads with its space program. The agency was in the midst of conducting a variety of flight tests on the shuttle *Discovery* to assess its aerodynamic abilities and structural integrity. If all the tests on the shuttle went well, NASA would need new astronauts, including scientists, to train for future missions. The time was ripe for women to be accepted into the program.

Until Sally saw NASA's advertisement, she had never considered becoming an astronaut. "I honestly never thought about doing it," she said. "It never even occurred to me. I thought I was going to get a job . . . doing research in free-electron laser physics and then work at a university doing research and teaching. That's what physicists do." Even though Sally knew nothing about the position being offered, she didn't hesitate for a moment in applying for the job. One thing was certain — there would be plenty of competition. NASA was swamped with over eight thousand applicants, more than fifteen hundred of them women, for the "astronaut class of 1978." Among them was another graduate student — a tall, red-haired astronomer named Steven Hawley, who would later become Sally's husband.

This was the first time in ten years that NASA would be hiring astronauts. NASA had begun developing the space shuttle shortly after the first manned moon landing in 1969. Now, eight years later, they were ready to form the eighth group of astronauts for the new space shuttle program.

What kind of qualifications was NASA looking for?

Teamwork. Poise and self-confidence. A cool, precise, and analytical mind. Expertise in a given field. These astronauts would have to be capable of conducting scientific experiments at zero gravity, monitoring the spacecraft's sophisticated, computerized equipment, and performing numerous other complicated and demanding tasks. All of this fascinated Sally, and also played to her strengths. Still, she knew this would be a challenge like none she had ever had.

Sally's graduate work in physics at Stanford had included research in X-ray astronomy — the study of radiation emitted by stars — as well as experiments with free-electron lasers and investigations into the theoretical behavior of free electrons in a magnetic field. At a time when relatively few women entered the fields of science and engineering, Sally was a respected member of the physics department. Jim Eckstein, one of her physics colleagues, said of Sally, "Everyone thought she was very good. But she was one of us. She quietly handled more than any of us there, and she never made a big deal about being a woman in the physics department."

With Sally's extensive science background, athletic prowess, scholastic achievement, and reputation as a team player she was just the type of astronaut candidate that NASA was seeking. Sally also met the other requirements NASA requested in their public announcement. She was thirty-two years old, eight years under the maximum age of forty, and her height of five feet, five inches was within the required sixty to seventy-four inches. Her eyesight fell within the 20/100 and correctable to 20/20 range, and she was in excellent physical condition.

Sally applied as a mission specialist, a job that required astronauts to have some biological science and at least three years of working experience in their field of study. Sally, aiming high as usual, sent in her completed application with health records and references. In October 1977 she was selected as one of 208 finalists. Sally was hopeful, but not overconfident. She was one step closer to her goal. Sally flew to the Lyndon B. Johnson Space Center outside Houston, Texas, for a detailed personal interview with the astronaut selection committee. She met with two psychiatrists, a team of medical doctors, and several NASA officials. Normally a cool and collected person, Sally remembers having a few butterflies in her stomach during the psychological interviews. "We saw two psychiatrists for about forty-five minutes each," she later commented. "One of them was generally exactly what I had always pictured from a psychiatrist — [a person] who showed you the comfortable chair . . . and then asked you how you felt about your sister. Then the other . . . was sort of the bad guy psychiatrist, who tried to rattle you."

Other parts of the interview were conducted by another group from the astronaut selection committee. They asked Sally exhaustive questions about her background, education, hobbies, research, and even her politics. The astronaut candidates were also put into what was called the "crystal rescue sphere." This was a little round compartment made of fabric; it seated one person and was designed so that an astronaut could transport a fellow crew member from one space vehicle to another if there were not enough pressurized space suits available for each person.

Part of the object of these tests was to subject the applicants to stressful situations and then examine his or her reactions. NASA's battery of tests and questions was also designed to find those who were the most serious about the jobs at hand. The agency was interested in scientists who were willing to devote their entire careers to the space program. For the scientists, many of whom were used to doing their own research, this represented a significant change in orientation. NASA had to impress upon them the kind of commitment they were expected to give. Thus the interviewers emphasized the long, hard hours of tedious mental and physical work that astronauts had to endure, and told the candidates that the job was not as romantic and glamorous as some might think.

NASA flight surgeons, meanwhile, gave Sally a physical exam, which lasted about two hours and included basic tests to check her hearing and sight as well as a treadmill test to check her heart. Finally, her review completed, she returned to Stanford University to finish work on her doctorate and await the decision, which NASA officials said would come by early December 1977.

On the morning of January 16, 1978, Sally received a telephone call from NASA official George Abbey: "Well, we've got a job here for you if you are still interested in taking it." Sally replied with an emphatic "Yes sir!" and she was on her way.

"I had no idea — which is amazing to me," Sally said in an interview some time later. "It didn't even occur to me that I'd get accepted. I thought I'd get into the last group." She had been selected as one of the thirty-five members of the astronaut class of 1978. Six of the new class members would be women, which meant that NASA had ushered in a new era of equality for women in space research.

Not surprisingly, Sally's father was elated at the news. "I think my Dad was kind of happy when I became an astronaut," Sally said with a laugh. "Before I joined NASA, I was in theoretical astrophysics. Astronaut was a concept he understood."

Before Sally started her astronaut training, she completed her doctorate degree in astrophysics. She was now, officially, Dr. Sally Ride, astrophysicist, a scientist who had the education, skills, and ambition to make the NASA team.

4

The Astronaut Training Program

IN 1979 SALLY Ride reported to the NASA-Johnson Space Center in Clear Lake City, a small community about twenty-five miles south of Houston, Texas, to become a member of the eighth class of astronauts. Very excited and just a bit nervous, Sally was also curious about the other thirty-four people reporting for duty.

She was especially eager to meet the five other women who had been selected for the program: Margaret Rhea Seddon, Kathryn Sullivan, Judith Resnik, Anna Fisher, and Shannon Lucid. No one knew who would be chosen to go up into space, but all the astronauts — men and women — would train and practice together.

Even though the space shuttles were launched from Cape Kennedy, Florida, the orbiters were guided into and out of space by technicians at the NASA-Johnson Space Center in Texas. This group of technicians, called Mission Control, was housed in just one of many buildings at the Johnson Space Center. Many of the other structures were used for astronaut training. As a whole the modern complex, with its grassy lawns and its network of sidewalks alongside the buildings, reminded Sally of a college campus. Over twenty thousand people worked in the center and crowds of tourists could be seen throughout the complex virtually every day.

Sally was soon to be called an astronaut candidate. All the new trainees were considered candidates until they had completed a one-year initiation and training period. In fact; once in the corps, the new members didn't even consider themselves astronauts until they had actually flown in space. Nevertheless, Sally and the other candidates had too much to do to worry about job titles.

Sally's research had interested NASA because free-electron lasers were thought to be a way of sending energy in space. NASA hoped that someday Sally's work could be used to develop a method of sending energy from a space station back to Earth. Sally had also investigated how electrons might behave when they are in a magnetic field. Thus it was natural for her to think that someday she would be able to perform her research in space, to see and study the earth, moon, planets, and stars from the unique vantage point of space. But first she had to pass her astronaut training courses.

To survive in space, astronauts have to command a broad array of knowledge well beyond their given specialty. First and foremost they have to know everything there is to know about their spacecraft, and so the candidates learned about the shuttle, from what it took to plan a mission to what the ground crew did while the craft was aloft. They learned about the different shuttle computer, mechanical, and electrical systems, much as a doctor might learn about different systems in the body. Each system's structure and the way it operated on the shuttle had to be understood and memorized. If there should ever be an emergency, the astronauts would have to recall this information immediately, without pausing to think.

Sally and the other astronaut candidates were taught how to handle a system failure, and how each system interfaces with the other. They were trained to take over for each other in case a crew member became sick or was unable to perform his or her job effectively. They learned about the functions of Mission Control, NASA operating procedures, first aid, aircraft design, astronomy, geology, engineering, survival techniques for air and water, how to eat, sleep, and exercise in space, plus hundreds of other complex functions and details required to operate the shuttle.

After completing the one-year training course, Sally would be able to parachute through trees and electrical wires, maneuver in a weightless environment, and survive being ejected from a

T-38 jet trainer. And since there was a good chance that they would become heroes to millions of Americans, they were taught how to speak well in public and how to answer the many questions that would come their way.

The first few weeks were as physically and mentally demanding as anything Sally had ever experienced. She awoke early in the morning, ran several miles each day, lifted weights, and conscientiously practiced what she had been taught. Sally didn't just read about survival techniques; she and the others were given real-life training — parachuting, being ejected from an aircraft, scuba diving in cold water, being plucked up from a raft adrift in the ocean by a noisy, wave-making helicopter, and more.

One particular training exercise, called "drop and drag," demanded every ounce of Sally's strength and endurance. Outfitted in her space suit and strapped into an open parachute, Sally was dropped from a moving boat and then had to release herself from the parachute harness while being dragged on her stomach and back. Once free of the parachute, Sally had to swim to a dock at the water training site.

There were other emergency escape techniques to practice as well. She learned what to expect if she had to jump from a moving aircraft and float safely to Earth suspended in a simulated parachute harness. To simulate ejection during an emergency escape from the shuttle, the astronaut trainee had to sit strapped into a seat

Astronauts must go through a year or more of grueling training before they're considered ready to go up in space. Left: Astronaut Ride boards the T-38 jet trainer for some high-speed flying practice. Below: Sally performs a demanding cold-water survival exercise.

that slides along a track at a very high speed —
and be catapulted off at some unexpected mo-
ment. The training for these skills took place at
Turkey Point, near Homestead Air Force Base in
Florida, and at Vance Air Force Base in
Oklahoma.

Complementing this rigorous survival training
was a full slate of classroom learning, with long
lectures in computer science, engineering, and
math. Fortunately, despite the barrage of information and new material, Sally found that her
fascination with astronomy continued to grow.
The candidates were also shown videotapes of
previous crews being questioned following the
completion of a mission, giving them firsthand
knowledge about the kinds of problems that
came up and of what they might expect in general. Sally began to feel that everything she had
learned and experienced was coming together,
and her excitement and anticipation of actually
going up in space soared with each passing day.

Finally, in August 1979 Sally and the other astronaut candidates completed their training and
evaluation period, making them eligible for assignment as a mission specialist on future space
shuttle crews. Sally and the other five female
candidates became the first women to cross into
NASA's all-male astronaut corps.

Some of the men who had been in the program
for years were slow to change their ideas about
women astronauts. Alan Bean, an Apollo and
Skylab veteran, was responsible for training can-

didates when this first class of women arrived. At first, he wasn't pleased with the idea. Nevertheless, he later admitted, "At first I imagined they were just individuals trying to do a man's job. I was proven wrong. . . . Females intuitively understand astronaut skills. They perform the mental and physical tasks as well as men."

The year of training was designed to change the way the astronaut candidates acted, reacted, and thought — and it did. The candidate's typical work week consisted of more than sixty hours that were both mentally and physically exhausting. Why did the astronauts endure so much hardship to become part of NASA's team? They certainly didn't stay with NASA for the starting salary of about twenty-five thousand dollars per year. They couldn't have been motivated by dreams of personal glory, since their achievements in space would be seen as a team effort. This work clearly had to be something to which they were very strongly committed. What could this something be? Like many other explorers, the astronaut candidates were remarkable people who were challenged by the unknown. For Sally Ride, there was also her enchantment with stars, a lifelong love of science, and her ambition to fly in space.

One of Sally's more surprising discoveries was that, like almost any job, there were boring and tedious aspects to preparing to become an astronaut. "People think we're tossed into centrifuges, dunked into water, and thrown out of planes in this work," Sally said. "But it's not always exciting. We sit behind desks and go to meetings mostly."

Even though Sally had finally become an astronaut by successfully completing her training, she was far from ready to fly in the space shuttle. She still needed actual experience working in and around the shuttle before she would be assigned to a flight. There was also a long list of older astronauts waiting for the chance to go up in space. Some had been waiting as long as fourteen years, and the new astronauts would have to wait their turn. While they waited, each astronaut was given a technical assignment which could last from two to five years. All of the new astronauts were also assigned to the Shuttle Avionics Integration Laboratory. SAIL was housed in a large, noisy building in which computers and machinery ran twenty-four hours a day. Groups of fifteen astronauts worked with SAIL personnel in both day and night shifts, examining the wires and circuits that made up the shuttle, testing procedures, and even simulating electrical problems. Sally and the other astronauts assigned to SAIL spent eight-hour shifts working in a duplicate of the shuttle cockpit. Monitors displayed various simulated problems and the astronauts' reactions were videotaped. It was hoped that the long hours spent working at SAIL would give Sally and the others the quick reflexes needed to effectively complete complex operations and respond immediately to a variety of situations. As Sally commented, "We have a simulator in Houston that's very good . . . they turn you on your back and shake you and vibrate you and pump noise in, so that it's very realistic."

Another requirement for all astronauts was to spend time flying in a T-38 training jet, the plane that accompanies the shuttle as it drops into the Earth's atmosphere upon its return from space. This flight training was especially important to new astronauts such as Sally who had had no experience flying a plane. Sally learned to read the cockpit controls and displays and felt what it was like to travel at high speeds in a confined space. During the long hours that Sally practiced in the T-38 chase aircraft, she also experienced the unpleasant sensation of the pull of gravity and the changes in perception that came from flying fast maneuvers. Sally thrived on the challenge and excitement. Indeed, she liked this flying experience so much that she continued her flight training and earned a pilot's license.

One of the biggest problems an astronaut faces in space is weightlessness, or zero gravity. Since weightlessness cannot be duplicated on Earth, training for weightlessness on Earth presented special difficulties. NASA experts found a few solutions to the problem. Sally and the other astronauts spent many long hours snorkeling in a "neutral buoyancy chamber" filled with water. It wasn't a perfect replica of what she would experience in space, but it was good enough for practice. The astronauts also practiced getting into their space suits, washing their hands, administering first aid, using a zero-gravity bathroom, and testing equipment for future spaceflights.

NASA engineers also devised a training procedure using a KC-135 military aircraft. The KC-135 aircraft would climb to a high altitude and then suddenly reverse direction and dive toward the Earth in a large arc. While flying at the upper half of the arc, the passengers were able to experience zero gravity for thirty seconds. Since anything not strapped down during those thirty seconds would float, the plane's interior had to be padded. As the plane completed the lower half of the arc, everything, and everyone, would thud to the floor. Again it wasn't perfect, but it was good enough. The astronauts nicknamed the KC-135 the "vomit comet" because some of them felt nauseated while the plane was simulating zero gravity. This unpleasant feeling in weightlessness is called space sickness. Luckily, not all astronauts are affected by space sickness; scientists are busy trying to figure out how to prevent it in those who are. In answer to a question about space sickness after her second shuttle flight, Sally commented, "That's very individualistic. About 40 to 50 percent of the astronauts do get sick and 50 to 60 percent don't, and I was one of the lucky ones. The symptoms range anywhere from classic motion sickness, which is a small percentage, to disorientation, where something just feels wrong. They feel like they're hanging from the ceiling and the earth is in the wrong place. Some people say they almost feel drugged, they're so tired. Generally it lasts about two days. . . . There have been three women up now, and interestingly enough, none of them have gotten sick . . . just putting my 'plug' in."

On April 14, 1981, Sally rode in the back of a T-38 chase plane during the descent of the space shuttle *Columbia*'s maiden flight. As the shuttle began its glide down to Edwards Air Force Base in California's Mojave Desert, Sally felt as if she were inside the graceful, soaring spacecraft. Sally could see that there appeared to be no serious damage to the surface of the orbiter. She provided weather information to the crew of the *Columbia* and also took pictures. When the shuttle softly touched ground, the astronauts, Mission Control, and millions of Americans breathed a sigh of relief. The first space shuttle mission had been a success. Sally Ride was one step closer to flying in space.

Sally was soon given a technical assignment — to help with the development, by Canadian scientists in Toronto, of a mechanical arm. This remote manipulator system, or RMS, was designed to lift satellites in and out of the shuttle's cargo bay while in orbit. The long robotic arm that the astronauts actually used to do the hoisting was to be operated from inside the shuttle. Sally and fellow astronaut John Fabian both trained on the mechanical arm and became experts on its use.

During the second and third shuttle flights, Sally was the first woman to work as a "capcom," or capsule communicator. While the term was a carryover from the old Mercury space program, the job was still critical in shuttle flights. The amount of time that the capcom had to talk to the shuttle crew was limited because the shuttle was within radio range of a tracking station for only a small amount of time. As a result, only one person on earth was assigned the task of talking with the crew in space at any one particular time.

Sally sat with the flight director at Mission Control and listened to discussions that the ground crew were having, especially when there were questions or problems. She then spoke to the astronauts, telling them what information they needed and giving them the flight director's instructions. At one point the astronauts aboard the shuttle's second flight told Sally of magnificent views of several sunrises each day. "The sun comes up like thunder and it sets just as fast," one astronaut reported. "And no sunrise or sunset is ever the same."

"Sounds good," Sally responded with eagerness. "When do I get my turn?"

That was in November 1981. What Sally Ride didn't know was that her turn was not far away. Her superiors, especially Captain Robert Crippen, were so impressed with her ability to perform under pressure that it was here that she became a strong contender as a crew member for the seventh shuttle flight. In addition to her physical and mental abilities, something else about Sally impressed NASA officials. Her enthusiasm and enjoyment of flying was akin to that of a special breed, the daring test pilot. As it turned out, she was the first astronaut chosen who did not have a military or test pilot background.

5

Assignment: Mission Specialist

IN APRIL 1982 NASA finished the selection process for the seventh space shuttle flight and released the crew list: Captain Robert Crippen, Commander; Captain Frederick Hauck, Pilot; Colonel John Fabian, Mission Specialist; Dr. Norman Thagard, Mission Specialist; and Dr. Sally Ride, Mission Specialist. It was the first time that such a large crew had been selected. It also included the first doctor to travel in space and the first American woman astronaut — Dr. Sally Ride.

George W. S. Abbey, NASA's director of operation, was asked why Sally was chosen to fly on the seventh shuttle mission. He responded that Sally showed a strong ability to solve difficult engineering problems. Perhaps even more important, he added, she was a "team player" who worked well with the other astronauts. "Sally

Ride is smart in a very special way," NASA's director continued. "You get people who can sit in the lab and think like an Einstein, but they can't do anything with it. Sally can get everything she knows together and bring it to bear where you need it."

Sally's wildest dream had come true. Her mother, Joyce Ride, said, "Sally is normally very cool, very low-key. But when she called to tell me the news, she was bordering on breathless."

However, Sally didn't welcome all the public attention given to her status as the first woman to be chosen for a flight crew. "I didn't come into the space program to be the first woman in space," she said at the time. Ride viewed herself as a mission specialist and as a scientist who also happened to be a woman. Whenever possible, she tried to minimize the attention she was getting by being selected as part of the shuttle crew. Still, the press seemed to ignore all other aspects of the mission, and Ride found herself overwhelmed with questions. Would she wear a brassiere in orbit? Was she afraid that the flight might damage her reproductive organs?

Sally felt that people should be more interested in the scientific nature of the shuttle mission. She marveled at how she could be of such interest when each shuttle trip included important scientific tasks never before attempted. Besides, Sally Ride was not the first woman to fly in space. A Soviet woman had flown in space twenty years earlier! In August 1982, just four months after the announcement that Sally Ride would be a *Challenger* crew member, the Soviet Union sent a second woman into space.

Once, when asked about all the attention, Sally remarked, "It's too bad this is such a big deal. It's too bad our society isn't further along. . . . It's time that we get away from that and it's time that people realize that women in this country can do any job that they want to do."

Sally was a private person who kept her feelings to herself. Therefore, she sometimes became annoyed with the press for asking her personal questions they didn't ask of male astronauts. Women astronauts, Sally felt, should be treated no differently than males. She had not expected to be treated differently when she had competed in boys' sports in her childhood, and she didn't expect special treatment now. It seemed perfectly natural to Sally that a woman should be chosen for an American space shuttle flight. After all, one of the goals of the shuttle program in the first part of the 1980s had been to open space travel to all people. The very job that Sally was to perform, that of mission specialist, was part of this new policy.

Sally's feelings were backed up by Captain Crippen. Though he had once joked that Sally was "the prettiest member of the crew," he later responded with some vehemence to a reporter who told him that Texans did not think a woman could do an astronaut's job. Crippen, a native Texan, replied: "I've found that all the women

43

astronauts have carried their share and more. We work together as a unit, but the fact that one is male and one is female, I haven't found made one bit of difference."

The other male astronauts also treated Sally as "one of the guys." When she herself was asked if she was treated differently, she quipped, "Crip [Captain Crippen] doesn't even open doors for me anymore."

Still, there were many unanswered questions, particularly medical ones, about how women would perform in space. Tests showed that women are more susceptible to the "bends," a hazardous condition that can afflict deep-sea divers if they surface too quickly. Could that problem affect women astronauts on space walks?

Another concern was the calcium loss in bones that was known to occur from prolonged weightlessness. Scientists had studied calcium loss in men on previous long-term spaceflights, and were able to estimate that, for example, on a one-year flight to Mars an astronaut could lose one-fourth of his or her total body calcium. It is also known that women naturally lose calcium more rapidly than men. But since the mission was scheduled to last only six days, Sally Ride was considered to be free of risk.

Sally's job on the upcoming shuttle flight would be to operate the fifty-foot-long remote-controlled crane, or robotic arm (RMS), that she and John Fabian had helped develop. Though they had traveled to Canada, where the device was being built, to consult with engineers about the design, they still had much to learn about its operation. The robotic arm had been successfully tested in space during the space shuttle's second flight, but it was to be put into actual use on the seventh flight. Sally and John had to know how to manipulate the arm to pick up satellites, lift them out of the shuttle's cargo area, and place them into space. They also had to learn how to use the arm to retrieve satellites from space and return them to the cargo bay. Accuracy, timing, and hand-to-eye coordination were essential, and the two mission specialists spent many hours practicing maneuvers during the year of training for the shuttle flight.

Sally had become fascinated with the robotic arm and practiced constantly, as if she were in training for a tennis match. She repeated the procedures for picking up small instruments and large, cumbersome objects as well. "It got to be as natural as using tweezers on a noodle," Sally recalled. "I began to think that all there was to being an astronaut was launching an arm." Or, as NASA official Carolyn Huntoon added, "She took to working that RMS like a duck takes to water, and really got some good vibes from all the people in that area. . . . It was designed for pilots, but she was doing better than the pilots were doing very early on."

Besides operating the robotic arm, Sally was designated to perform several other important jobs in space. She was to serve as the flight engineer, assisting the commander and pilot during ascent, reentry, and landing. If an unexpected situation occurred during the flight, Sally's role as flight engineer would be to suggest corrective procedures. If all went well, her duties would be to monitor the massive amount of data from the computers on the flight control panel.

The crew members spent the next year preparing for their mission, in the process becoming a close-knit family. They shared an office at the NASA-Johnson Space Center in Houston and spent most of their working time together. They practiced in the shuttle cockpit simulator, sometimes for as long as fifty hours or more at a stretch. They rehearsed all parts of the mission and for possible emergencies.

On July 24, 1982, Sally Ride and Dr. Steven Hawley were married, making them the first pair of astronauts to do so. Hawley was from Kansas and had a Ph.D. in astronomy from the University of California. The couple had met in 1978 when they were both in training for the space program at NASA. Sally Ride flew her own plane to the wedding in Kansas, making good use of her pilot's license. They moved into a comfortable house outside the Johnson Space Center. Their home was filled with space age mementos, including NASA posters, shuttle dishware, and even a large photograph of the 1968 moon landing.

6

Six Days in Space

SPACE SHUTTLE *CHALLENGER* has delivered to space the largest human payload of all time — four men, one woman." This was one of Mission Control's first announcements after the shuttle Challenger reached its orbit 184 miles above Earth on June 18, 1983.

In the *Challenger*'s cargo bay were four primary payloads: two communication satellites to be placed in orbit and two satellites that were to return to earth with the shuttle after tests and experiments were conducted. Also on board were: seven "Getaway Special" canisters housing twenty-two experiments designed by high school and university students, a private company, and two government agencies; classified Air Force experiments on cosmic radiation, which were to be performed for NASA; and a group of projects called the Small Self-Contained Pay-

47

loads Program. These last were selected from many individual scientific research projects and were lofted into space for fees ranging from one to three thousand dollars. One of them, devised by high school students from Camden, New Jersey, intended to study the effects of gravity and weightlessness on a colony of 150 carpenter ants. The ants' activities were to be monitored by the crew through the use of a television screen.

With the tension of the launch behind them, Mission Control was eager to hear how the crew felt. Roy Bridges, another NASA astronaut, served as Mission Control communicator on the first day of the flight. "How is it up there?" he asked.

"Have you ever been to Disneyland?" Sally Ride asked Roy.

"Affirmative," Roy replied, using formal radio language for "yes."

"This is definitely an 'E' ticket," replied Sally, expressing her enjoyment in a way that her earthbound astronaut friend could surely understand. Disneyland's most fantastic rides, such as the submarine voyage or mountain roller coaster, required an "E" ticket for admission.

In the vacuum of space, temperatures range from plus 200 degrees Celsius on a surface exposed to direct sunlight to minus 200 degrees Celsius on a surface shielded from the sun's rays. But in the controlled atmosphere of the *Challenger,* the temperature inside the cabin was kept between 70 and 75 degrees Fahrenheit, enabling the astronauts to breathe comfortably and wear light clothing — short-sleeved cotton T-shirts, pants, or gym shorts.

Sally gazed with awe out the space shuttle's window. The sensational panorama of the glistening blue oceans and radiant orange deserts against the black background of space was even more magnificent than she had imagined. Although they weren't far enough away from Earth to see the entire planet at once, she could see Oregon when they flew over Los Angeles, and New York as they flew over Florida.

The tallest skyscrapers, giant bridges, and sprawling cities looked like tiny specks on a large blue marble. Sally could see twinkling city lights, dust storms blowing over deserts, and electrical storms raging over oceans. The Great Lakes looked like little silver-blue water droplets. The towering mountains of the Himalayas in Asia looked like little pieces of crinkled aluminum foil. Dramatic clusters of clouds floated effortlessly over Africa.

Then Sally looked outward, beyond the Earth, to the black velvet cloak of space. She could see countless pinpoints of light where billions of stars, planets, and galaxies were shining just as they had since the beginning of time. These sights made Sally feel very special and she would remember them and the feelings they evoked for the rest of her life.

While the sights from space were breathtaking, every moment the astronauts spent in orbit was precious, so they kept very busy almost all of the time. Their workload was heavy, and the time they had to complete various tasks — including "housekeeping activities" such as eating, sleeping, washing, and cleaning up in the shuttle cabin — was precisely regulated according to a strict timetable preplanned by NASA engineers. Each of the astronauts had a list of jobs that were to be completed, and all the shuttle's systems needed to be checked out.

Now that the Challenger was safely in orbit — it would make one complete revolution of the Earth every ninety minutes — the next order of business was to release the first payload into orbit. This was the Canadian-sponsored Anik C-2 communications satellite, which would provide North America's first direct satellite-to-home pay-TV service. The fact that NASA was being used to launch this satellite represented the agency's plan for the space shuttle to do commercial work in space in addition to its conventional role of running its own scientific experiments. If successful, NASA stood to earn several million dollars in addition to proving its viability as a commercial enterprise.

As the *Challenger* approached its seventh orbit, mission specialists Sally Ride and John Fabian stood in front of a control panel at the rear of the flight deck. From here Sally could see the satellite launching platform through two small windows and a television monitor. She looked at the controls she knew so well from her simulations and prepared to use the robotic arm of the remote manipulator system. Commander Crippen and Pilot Hauck opened the cargo doors and started the procedure to position the shuttle for deployment of the satellite. The scientists and computers in Houston worked hard to supply the astronauts with all the necessary data they needed to complete the careful maneuvering required to get the shuttle into the correct position to launch the satellite. When this had been done, a signal was given and the satellite started to spin.

"It sure is fun," Sally told Mission Control when she began to operate the satellite deployment controls. She watched the satellite spin in place at fifty revolutions per minute. It would continue spinning while being ejected and while operating; this helped to keep it stabilized and in its correct orbit. If it didn't rotate properly, there was a chance that its delicate solar panels would melt if it became exposed to unfiltered sunlight.

The crew waited twenty minutes until the systems for launching the precious payload were ready and the computers sent the signal to release the satellite. Finally, about nine and one-half hours into the mission and toward the end of their first day in space, the large satellite was spring ejected from its spinning platform in the cargo bay. The astronauts felt a slight tremor of the spacecraft as the satellite spun free of the cargo bay. The complicated device was operating as planned.

As shuttle rockets fired to move the orbiter away from the spinning satellite, the crew gave a sigh of relief. Looking toward the cargo bay, Sally could see the bright glow of the firing at the aft end of the spacecraft. To complete the launching procedure, a rocket motor attached to the payload was fired to begin the satellite's four-day ascent to its orbit, 22,300 miles above the Earth's equator. As the Anik C-2 communication satellite floated away from the *Challenger,* Sally felt a great sense of accomplishment. She had completed her first primary mission in space. The crew was jubilant that their first day's work had gone as planned, and Mission Control declared the satellite launch a success.

On her first day in space, Sally decided that being weightless was a wonderful sensation. She drifted without effort about the flight deck and experienced without ill effect the quirky sensation that being upside down was no different than being right side up. Moreover, she enjoyed the fact that the slightest touch could start her body floating across the room! In order to stop moving, she had simply to grasp on to something that was anchored in place. This part of the job seemed more like a fun game than work!

Astronauts did, however, have to make certain adjustments to being weightless, especially when it came to eating, sleeping, or using the lavatory. Eating and preparing meals aboard the space shuttle presented NASA engineers and astronauts with a true challenge. Floating food particles or water

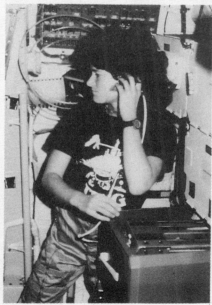

Aboard the shuttle Challenger. Above: This view through the shuttle's open cargo bay shows the Earth in the background and the 50-foot-long robotic arm of the remote manipulator system, used to deploy and retrieve satellites from inside the orbiter. Left: Though temperatures in space range from +200 to −200 degrees Celsius, inside the zero-gravity cabin Sally and her colleagues wore light clothing as they worked.

droplets could cause the shuttle's sensitive and complex equipment to malfunction. Even though Sally could sit on the ceiling of the shuttle and munch on a handful of nuts if she wanted to, when it came to fixing a meal, more careful planning was required. Meals were prepared by two mission specialists and served in the *Challenger*'s galley, or kitchen, which was located in the dining area on the lower level of the cabin. Although they were orbiting high above the planet, the shuttle astronauts had a comprehensive menu which boasted twenty beverages and over seventy-five varieties of food.

A typical menu for a particular day included: scrambled eggs, peaches, and orange drink for breakfast; frankfurters, almond crunch bar, bananas, and apple drink for lunch; and shrimp cocktail, beef steak, broccoli au gratin, grape drink, and butterscotch pudding for dinner. Sticky foods such as macaroni and cheese or peanut butter were easier to keep on a spoon. While the average menu included many more items to choose from, it was repeated every six days.

Their main eating utensils were a spoon and a pair of scissors, used to snip open the pouches of freeze-dried and vacuum-packed food. The dehydrated food was prepared by injecting water into its plastic bag and waiting a few minutes for it to take form. Other foods, such as ham or stewed tomatoes, were precooked before the launch and stored until they were heated in a microwave or convection oven. Pears and other fruits were

54

freeze-dried; nuts and cookies were kept in their natural form.

Meal preparation took about twenty minutes. Serving involved attaching the food pouches to trays using Velcro; Sally and the other astronauts could then strap the trays to their legs, mount the trays on the shuttle's cabin walls, or eat at a special table with restraints. The astronauts often ate picnic-fashion while floating in midair, and usually ate together so they could relax a bit from their busy schedules.

Cleaning up after meals was also an important task. NASA scientists found that microbes, if left unchecked in the cabin could multiply extremely fast in the weightless environment of the shuttle, causing illness among the crew. To avoid this potential hazard, Sally and the other astronauts had to make sure that their eating utensils were wiped clean and sterilized before their next meal. Many of the items used during a meal were disposed of in garbage bags and stored until the shuttle returned to Earth. There was plenty of water on board for washing and drinking because it was a by-product of the fuel cells that produced the orbiter's electricity.

The shuttle was equipped with personal hygiene stations where Sally and the other astronauts could wash their hands, brush their teeth, and use the lavatory. The *Challenger*'s toilets were much like those on a commercial airplane, except that they had foot restraints and seat belts, and the biowastes were swept away by a fast moving current of air instead of water. Biowastes were not dumped overboard but instead were vacuum-dried, stored, removed after landing, and often tested for mineral loss and other important medical information concerning an astronaut's metabolism in space.

As on Earth, astronauts need a good night's sleep to properly do a full day's work. There were two basic sleeping systems available on the shuttle: the cocoon-type sleeping bag and the rigid sleep station. There were four rigid sleep stations in the shuttle's cabin, which included a sound suppression blanket, sheets with restraints, an overhead light, a communication station, and a pillow. The sleeping bags were made of a perforated material which made for better temperature control. Because of the many sunrises and sunsets over the eight-hour period that the astronauts slept, they had to wear eye masks to shut out the light. They also used earplugs to block out background noise from radio static and automatic "burns" by the shuttle maneuvering rockets.

Sally chose the sleeping bag for her bed rather than the rigid sleep station. Since no way is up in a zero-gravity environment, Sally situated herself in her bag as if she were standing up. She zipped herself into the bag, put on her cassette tape player and earphones, and drifted up to the flight deck ceiling next to a window. Before covering her eyes with the sleeping mask she looked out the window while listening to music and watched Earth floating silently below. Everything was so hauntingly beautiful! Sally could barely contain her excitement. She didn't want to waste a minute of time, certainly not on sleep, but she knew she had to be alert for the next day's work. Reluctantly, Sally covered her eyes and tried to push all the wondrous things that had happened to her that day from her mind so she could fall asleep.

On the mission's second day in space, a smaller satellite was launched and fired into a 22,300-mile-high orbit. This satellite would serve Indonesia and other Southeast Asian countries by providing better telephone service. Using the same technique as she had the day before, Sally's confidence in operating the shuttle's sophisticated equipment quickly grew.

Sally and the crew next turned their attention to the other equipment, tasks, and experiments. The "Getaway" canisters had to be activated to test the effects of space on radish seed, germinating snowflowers, liquid mercury, and soldering operations. Two of the canisters had special technical features which included the first barometric switch and an automatic opening door. The Ku-band antenna, designed for ground communications through satellites in orbit, was checked out for motion and its ability to receive a signal. It worked properly and was declared ready for operation.

Sally's third day in space included a test of a procedure that would shorten the time an astronaut needed to prepare for a walk in space. As it was, an astronaut had to breathe pure oxygen for three hours before leaving the shuttle's controlled environment to work in space. The test, which was conducted while the astronauts slept, reduced the cabin pressure by controlling the mix of oxygen and nitrogen. The experiment worked, setting the stage for new procedures to be used in the future; on this mission no walk in space was scheduled.

One of the mission's few failures occurred on the third day, when one of the flight deck television screens designated to provide the pilot and commander with vital reentry information failed. For a short while, the astronauts considered trying to find the source of the malfunction and fix the television monitor, but they found that they could use the remaining two screens and still safely land the shuttle.

Sally's fourth day in space began with work on a pharmaceutical experiment for Mcdonnell Douglas. Meanwhile, other ongoing experiments were checked, the shuttle's systems were monitored, and, of course, there was lunch and a between-meals snack.

As on Earth, exercise is an important part of an astronaut's daily routine. In fact, it is even more vital in space because exercise counters the negative effects that weightlessness has on muscles, which tend to waste away the longer an astronaut stays in zero gravity. Since there was relatively little room to move about the shuttle's cabin, Sally and the other astronauts exercised on a treadmill device. They also played catch with a bag of jelly beans, a special gift from President Ronald Reagan, and played a frisbee-type game using, of all things, a cookie. By the end of day four, Sally and the other astronauts were getting very accustomed to their new home.

Throughout the flight, experiments were being performed on the astronauts themselves. Mission specialist Dr. Norman Thagard was in charge; he conducted tests on the entire crew, himself included, measuring fluid motion and pressure inside their heads, eye movements, visual perception, and more. With the new information that was collected, NASA scientists hoped to find a solution to Space Adaptation Syndrome, or space sickness, the troublesome illness that affects so many astronauts.

On the fifth day in space, Sally and Colonel John Fabian were scheduled to test the automatic satellite launch system by using the robotic arm to pick up, send into, and retrieve from space a West German satellite called SPAS, or Shuttle Pallet Satellite. The satellite, part of NASA's commercial endeavors, contained remote-sensing instruments and eleven experiments, among them microgravity experiments with metal alloys, heat pipes, and air pressure conveyors; a new instrument that, it was hoped, could control a spacecraft's position by using a remote-sensing scanner to pinpoint different kinds of terrain on earth; a mass spectrometer for monitoring gases in the shuttle's cargo bay and around the orbiter's jet thrusters; and a new method for calibrating solar cells.

The commercial concept behind the SPAS was that space for experiments on the pallet satellite would be sold to different customers. After their experiments were attached to the pallet, the satellite would be placed in the shuttle's cargo bay, launched into space, and, once in orbit, provided with power and computer processing. After the experiments were finished and the shuttle returned to earth, the satellite pallet would be removed, refitted with new experiments, and reused again and again.

Sally, preparing to operate the robotic arm, watched through the shuttle's rear windows and on the television monitor as commands were sent to activate the pallet satellite. Sally pressed the buttons to manipulate the spindly arm, and its clawlike hand gripped the satellite and pulled it out of the cargo bay, releasing it into space. Next, Commander Crippen and Pilot Hauck fired the shuttle's small jets to maneuver the spacecraft slightly below and ahead of the free-floating object. Using small yet precise jet firings, Crippen and Hauck moved the *Challenger* to within one thousand feet of the satellite, and then, with painstaking accuracy, closer and closer. Finally, they were near enough for Sally to use the robotic arm to grab the satellite. Happy with the success of the maneuver, Crippen quipped, "We pick up and deliver."

Over the next nine hours, Sally and the crew continued their experiments with the remote arm and the pallet satellite. They maneuvered around the satellite, approached it, grabbed it, and returned it to its position in the cargo bay, thus proving that the shuttle could safely interact with another spacecraft in orbit. All the while, Sally monitored the temperature inside SPAS. At one point she noticed that the internal temperature on the data processing unit of the pallet satellite was climbing higher and higher. If it overheated, she realized, the satellite would not be able to receive operation commands. Halfway through the experiment, Sally recognized that the temperature was too high. Something was definitely wrong — something had to be done immediately!

Sally quickly shut off many of the satellite's systems to allow them to cool down. Then she and the other mission specialists discussed alternatives that might solve the problem of the overheated satellite. They reasoned that since space has no atmosphere to block the sun's rays, objects in direct sunlight increase in temperature. Therefore, if they could maneuver the *Challenger* to block the sun's rays, perhaps the pallet satellite would cool down even more. This proved to be very difficult, even with all the practice maneuvering the crew had been doing. Finally, the commander and pilot were able to get the *Challenger* into the desired position. The problem was solved.

Sally retrieved the satellite for the last time at 1:40 P.M. and brought it back to its permanent place in the shuttle cargo bay. The robotic arm had been tested for almost nine hours and had operated flawlessly. The crew had also tested the shuttle's stability while the arm held a five-thousand-pound weight as the orbiter's jets were fired in short bursts. Sally was elated that things had gone so well. She took another glance at Earth, seeming so small in the black sea of darkness, and thought once again of the wonder and splendor of the past five days.

Sally's journey in space had included a lot of hard work, a tremendous amount of teamwork, and challenging adaptation to new living conditions. The satellite launchings and tests had been successful; indeed, the mission had accomplished 96 percent of its assigned goals. Sally crawled into her sleeping bag that last night in space feeling satisfied that she had performed her job as mission specialist with the high standards she had always set for herself. Though her part of the mission was complete, the journey was far from over. Sally and the other astronauts had one more important task to accomplish — returning safely to Earth.

Early on the morning of June 24, their sixth and last day in space, the crew of the *Challenger* began preparing for reentry into the Earth's atmosphere. The day before they had floated around collecting the books, pencils, and small equipment that had drifted against the ceiling or walls during the course of the flight. These objects would crash down upon them during reentry if they weren't stowed away in drawers. The astronaut's reclining seats, which were stored after the launch, were brought out and reattached to the floor of the flight deck. Sally and the other astronauts put on their space suits, boots, helmets, and life vests for the dangerous descent through the atmosphere and return to Earth. They were ready to go home.

Unlike previous shuttle flights, which landed in the California desert, this *Challenger* mission had planned to inaugurate a new three-mile-long runway that NASA had built at the Kennedy Space Center in Florida. This shuttle "first" would eliminate the cost and time it took to ferry the orbiter — mounted on top of a special 747 jet airliner — from the landing strip at Edwards Air Force Base in California back to its launching pad at the Kennedy Space Center. Unfortunately, owing to clouds, rain, and fog in Florida, NASA engineers made the decision to land at Edwards Air Force Base in California.

About five hours before the anticipated reentry, each astronaut began to drink at least four large glasses of liquid and to take salt pills to keep the fluids in their bodies. This was done to prevent their feeling thirsty and light-headed when gravity forced their bodily fluids to flow toward their legs as the Earth's gravitational pull increased.

Sally knew that her face had been a bit puffy during the past six days; this was because the fluids in her body hadn't been pulled toward her lower body, as normally happened on Earth. She also knew that, if necessary, she could inflate the antigravity pants on her space suit to keep the blood from settling in her legs, which would make her feel light-headed during reentry. As the astronauts strapped themselves into their seats and connected their helmets to the oxygen supply, Sally knew that this great adventure was about to come to its dramatic conclusion. The excitement of lift-off returned and her mind was filled with questions. What would reentry feel like? What would Earth look like as they got closer? Would the shuttle's special heat-resistant tiles protect them from the tremendous temperatures generated as the shuttle reentered Earth's atmosphere? What if something went wrong?

Sally's mind snapped back to the moment at hand as the shuttle's small engines fired and *Challenger* left its orbit on the ninety-seventh revolution. There was no turning back. In about thirty minutes, the orbiter would begin to reenter Earth's thinnest layer of atmosphere traveling at the incredible speed of about seventeen thousand miles per hour!

To slow down the spacecraft, the orbiter had to be turned around and the engines fired in the opposite direction to which they were traveling. This complicated braking maneuver was necessary to get the speed of the falling spacecraft down to about two hundred miles per hour. If it failed, *Challenger* and its occupants would burn up like a falling meteor in Earth's thickening atmosphere. As the spacecraft turned, Sally felt the speeding orbiter vibrate violently and heard the noise of the blasting engines. Steady and calm, she kept her gaze fastened on the computers and display terminals as they guided the shuttle back to Earth. The shuttle was then maneuvered to face frontward again and continued its journey homeward.

As the shuttle reentered Earth's atmosphere, it moved at such a speed that friction from air molecules heated up the outside of the spacecraft to about 2,700 degrees Fahrenheit. This created an orange glow of heat which surrounded the shuttle, much like the light of a large bonfire. Although she was safe inside, Sally knew that this intense heat blocked radio transmission from the shuttle to Mission Control. This meant that the *Challenger* crew had to rely on the on-board computers until the ship passed through this blackout zone.

When the shuttle came in contact with Earth's atmosphere, Sally and the other crew members heard an enormous rush of air — it was as if they had entered a vast wind tunnel. Gravity began to pull Sally down into her seat, and the orbiter vibrated as an airplacne might when experiencing turbulence. Although her body felt heavier and heavier, her spirits were high. By the time the shuttle received Mission Control's radio transmission there were only thirty minutes left until landing.

Commander Crippen switched off the automatic pilot and took manual control of the *Challenger,* which now maneuvered like a glider traveling at about two hundred miles per hour. After resuming contact with Mission Control, he asked them vital questions about wind velocity and direction and received any other critical information he needed to secure a safe landing. He then maneuvered the shuttle in wide sweeping turns, and the spacecraft took on the look of a giant eagle swooping down to its nest. The *Challenger,* which had begun its journey as a rocket, was ending it as a colossal bird.

As the shuttle glided down through the clear desert sky and approached the lake-bed runway, Sally saw the T-38 chase planes outside the *Challenger*'s window escorting them home. Finally, at 6:57 A.M. (PDT) the space shuttle *Challenger* touched the ground with a bounce; they had landed safely and on target. The tires screeched and the parachute opened, slowing the shuttle's speed on the long runway. Sally relished the feeling of solid ground beneath them — she and her fellow astronauts were home, safe and sound. Mission accomplished!

Sally's first flight in space had lasted six days, two hours, twenty-four minutes, and ten seconds. During this time, the *Challenger* made ninety-seven orbits around Earth and traveled a distance of 2.2 million miles.

Sally and the other astronauts now needed to adjust to Earth's gravity. Although they had all exercised during their six days in space, zero gravity had weakened their muscles. In particular, their hearts had not needed to pump blood up from their legs, against the pull of gravity; on Earth their hearts had to work harder and beat faster. The astronauts would also need time to regain their sense of balance. Sally knew that it would take about fifteen minutes to feel comfortable walking in a straight line and that she would probably feel wobbly for about half an hour.

Since the decision to land in California had been made four hours before the *Challenger* touched down, only a small crowd was on hand to meet Sally and the other crew members. Her family and numerous spectators were waiting for their return in Florida, and, though disappointed that they could not witness the landing, they were happy that the mission had been a success and that the astronauts had returned safely. President Reagan, too, had hoped to meet the crew in Florida but was forced to cancel his plans so as not to influence the *Challenger*'s landing site. He did speak to the astronauts on the phone that day, warmly congratulating them all. He had specific praise for Sally Ride. "You were," he asserted, "the best person for the job."

7

Becoming a Celebrity

ALTHOUGH SALLY RIDE didn't join NASA's space program to be the first American woman in space, her life changed considerably as a result of this accomplishment. Though she had been the subject of considerable media attention just before her maiden voyage into space, upon her return she was virtually deluged with requests for newspaper interviews, television appearances, and lectures. As a private person, such a high profile caused her some discomfort. Still, she acknowledged, "I've come to realize that I will be a role model."

This was especially so for children. Not surprisingly, many of the questions asked by curious youngsters revolved around routine activities. The following exchange is typical of the many Sally had.

"How do you take a shower or bath in space?"

"You don't." Sally responded. "We don't have a sink, bathtub, or shower because water would float in little blobs all over the cabin. . . . We have a water gun. . . . I wet a washcloth. . . then I use it with soap to wash my hands and face."

"What was the space food like?"

"The food they give us is really civilized. . . packages of cookies, peanuts, M&Ms," Sally answered. "All the drinks are dehydrated — iced tea, coffee, orange juice, apple juice. . . and the foods are like macaroni and cheese. . . . They're not great but they're also not that bad."

"How do you sleep in space?"

"Sleeping in space turns out to be easy." Sally explained. "Most people end up with a sleeping bag around them. . . just floating free. . . someplace in the middle of the cabin."

"What did you do with your free time in space?"

"It turns out that there is a fair amount of free time just after you get up and right before you go to bed," Sally responded. "You can spend a couple of hours at the window."

"How do you go to the bathroom on the space shuttle?"

Sally explained that it was similar to those on commercial airplanes, but joked, "A mistake you only make once is not turning on the air suction."

Between Sally's first and second flight into space, a span of about a year-and-a-half, she spent a considerable amount of time traveling around the country, talking about NASA and the shuttle program and receiving recognition for her accomplishments. She was an especially busy celebrity in July 1983, the month after the flight. At NASA headquarters in Houston, Texas, Sally was recognized by about seven hundred women government employees as a leader in the cause of equal opportunity for women in the workplace. She went to the Smithsonian National Air and Space Museum in Washington, D.C. to donate her space suit, attend a state dinner at the White House (along with the rest of her shuttle crew), and testify before the Senate Commerce, Science, and Transportation Committee. After this last appearance, Senator Bob Packard of Oregon presented her with a T-shirt reading, "A Woman's Place is Now in Space."

In August 1983 Sally and the shuttle crew were given "keys" to New York City by Mayor Edward Koch. They in turn presented him with a photo montage of the space shuttle. In October Sally was honored by the Girls Club of America at a reception hosted by *Ms.* magazine publisher Gloria Steinem and tennis star Billie Jean King. *Ms.* had honored Sally by putting her picture on the cover of its January 1983 issue. She took the magazine with her on the shuttle mission and later donated it to the National Air and Space Museum. At the Girl's Club reception, Sally was given the opportunity to run in the 1984 Olympic Torch relay and met with the organization's young members to discuss how her career as an astronaut developed.

In January 1984 Sally was invited to be a guest on "Sesame Street," the children's television program, where she gave a lesson on the letter "A" and helped Oscar the Grouch and his girlfriend, Grundgetta, get ready to take a televised ride into space. In June Sally was given the Jefferson Award from the American Institute of Public Service in a ceremony hosted by FBI Director William Webster.

Even with all her "obligations" to the public, Sally Ride continued to prepare for her second shuttle mission. She reviewed flight data and practiced fire fighting, water rescues, parachute jumps, robotic arm maneuvers, flight training, and all the other necessary procedures. After all, in October she was scheduled to return to space.

8

A Second Voyage into Space

ON OCTOBER 5, 1984, Sally Ride was strapped into her reclining seat, about to leave on her second space flight, the thirteenth successful launch in the shuttle program. She and six other astronauts, one of them a woman, formed the largest crew the shuttle program had seen to date.

Sally was much more confident now and felt that her training for the second mission had gone much more smoothly because she had known what to expect. She was also thrilled that she was no longer bombarded with questions about being a woman in space. She did speak to a group of reporters before the flight about the energy monitoring satellite that *Challenger* flight 41-G, as this mission was called, would be carrying into space. The satellite was designed to measure the amount of solar energy reflected into space by the Earth's atmosphere. Through these measurements, scientists hoped to learn more about the Earth's weather.

"We're going to leave that satellite in orbit to conduct its experiments and to look for things like the greenhouse effect," Sally explained to reporters. The greenhouse effect is the name given to the process by which carbon dioxide traps radiation and its heat in the Earth's atmosphere. The carbon dioxide acts much like the glass ceiling of a greenhouse; though transparent, it is able to hold in heat. Scientists are worried about the greenhouse effect because carbon dioxide levels in the earth's atmosphere are increasing and may be the cause of rising temperatures and other profound climactic changes all over the world.

Sally's thoughts were interrupted by a voice coming through her helmet intercom from Mission Control: "Launch minus seven minutes." As the astronauts and Mission Control did their final check to make sure that all systems were ready for lift-off, the walkway which attached the launch tower to the orbiter slowly pulled away from the shuttle. Sally and the others felt the craft vibrate as the power units began to whirl. When the seven astronauts lowered the visors of their helmets, they inhaled the oxygen being pumped into their helmets and waited as they heard a voice from Mission Control say, "Launch minus ten seconds."

Three seconds later the shuttle's three main engines ignited and the *Challenger* shuddered, trying to break free of the bolts which fastened it to the launchpad. What happened next would be up to the computers. If all the shuttle's systems worked properly, the computers would allow the launch to continue. If not, the computers would shut off the shuttle's main engines and the launch would be scrubbed. Finally, after what seemed like an eternity but in reality was only a matter of seconds, the shuttle's solid rocket boosters fired and the *Challenger* and its crew lifted off in a white cloud of steam, dense black smoke, and a stream of fire. After two minutes, the din of the booster rockets stopped — their fuel spent, they had burned out. Although the ride was now smooth and quiet, the launch engines continued to drive the shuttle higher, out of the atmosphere and toward the vacuum of space. Then, without warning, the intense force pushing against Sally disappeared and she fell forward in her seat; a safety harness held her in place. Sally knew what the feeling of lightness meant and, as she saw pencils and books floating about the shuttle's cabin, she realized that the launch had been a success.

On *Challenger* flight 41-G Sally was one of five mission specialists. Her primary task was to launch the weather satellite using the robotic arm. Kathryn Sullivan, a thirty-two-year-old geologist and expert in remote sensing equipment, would be the first American woman to walk in space. Remarkably, Sally and Kathryn had been in the same suburban elementary school in Los Angeles nearly twenty-seven years earlier; neither NASA nor the women themselves had had any idea that the two spacemates had been childhood classmates. As Sally recalled, "We were sitting around talking one day, and we discovered we were in the first grade at Havenhurst Elementary School at the same time."

The other crew members were Commander Robert Crippen; Pilot Jon McBride; Mission Specialist David Leestma, an aeronautical engineer; Payload Specialist Paul Scully-Power, an Australian oceanographer; and Payload Specialist Marc Garneau of Canada. Since this was a crowded flight, NASA officials stressed the need for "good housekeeping," so everyone did his or her part to keep the shuttle clean of small food particles that could have caused problems.

Even though the shuttle had had a trouble-free launch, the flight encountered a good number of technical difficulties. The earliest problem came on the first day of the mission, when Sally tried to launch the weather satellite. When she picked up the satellite from the cargo bay with the robotic arm, she discovered that the springs and hinges on the satellite's solar panels had frozen. These would have to thaw before the satellite could be launched. Remembering her previous mission's success at cooling the overheated satellite by using the shuttle craft as a shield from the rays of the sun, Sally figured that doing the opposite would work this time. She had Crippen and McBride maneuver the shuttle's cargo bay into direct sunlight, and after a brief exposure to the sun the hinges on the satellite began to thaw. The weather satellite was launched without further delay.

After a well-deserved evening meal, Sally got into her sleeping bag, positioned herself near a window on the flight deck, and again looked at the wonders of space before falling asleep. Eight hours later, as she awoke, she found herself a little confused. Was that the theme song from *Flashdance* she was hearing? It was; by space shuttle tradition Mission Control always supplied a musical wake-up call on each flight's first morning in space. She smiled because the crew had come up with some entertainment plans of its own. They had brought along a taped reply made especially for that morning: "The crew is temporarily out. Please leave your name and address so that we may return your call later on."

The upbeat nature of the crew was sorely tested over the following days, for the mission was plagued with more problems. An antenna malfunctioned, meaning that they had to store data from the imaging radar on tape, then turn the shuttle until the antenna pointed at a satellite so it could be transmitted to Earth. At one point a large radar panel would not fold up properly. Sally came to the rescue by using the robotic arm to grab the panel and gently squeeze it shut until the automatic latches caught.

Despite the problems, the live television pictures transmitted back to Earth showed a happy crew. "We're having a lot of fun," Sally smiled as she floated in front of the television camera. "We've had a few problems, but so far we've been pretty successful in almost everything we've tried to do."

On the fifth day the astronauts scheduled a news conference with journalists from around the world. Unfortunately, serious flooding in parts of Asia hindered phone connections with that part of the planet. As Mission Control struggled to get all the journalists on the phone, a recorded message was accidentally sent over space-to-ground communications — "If you need help, please hang up and try the operator" — prompting loud laughter from outer space.

The highlight of the mission, Kathryn Sullivan's space walk, was postponed for two days because of a malfunction in the air- conditioning. Finally, on October 11, astronauts Leestma and Sullivan left the shuttle to walk in space for a successful test of a satellite refueling technique.

After the mission objectives had been completed, it was time for *Challenger* to return to Earth. Despite Hurricane Josephine threatening troublesome weather, *Challenger* touched down on the Kennedy Space Center Shuttle Landing Facility runway shortly after midday on October 13. Even with all the technical difficulties this shuttle mission had experienced, the program as a

78

whole was succeeding at moving toward NASA's goal of making space travel available to everybody. The shuttle had a good safety record and NASA was eager to send up more commercial cargo to help cover some of the shuttle's costs.

Sally continued to make public appearances upon her second return from space. On November 3, 1984, she returned to Stanford University to speak to an excited group of students, faculty, and staff. After she showed the audience a spectacular film of her latest mission, the students and autograph seekers, mostly young women, gathered around her to ask questions.

"What was weightlessness like?" "Your body is a little bit confused by it the first day and it just takes a little while to adapt," Sally answered. "Everybody's first reaction is to just try and swim, and that doesn't work. . . but after a day or so, you figure that out, and physiologically you really don't feel any different. Your body adapts very quickly and really seems to enjoy it."

"What was it like just after you landed?"

"It feels much more difficult physiologically." Sally explained. "Your heart rate doubles. . . you turn your head and the whole room spins, your arms and legs feel very heavy, you feel like the book that you were just able to float in front of you now weighs three hundred pounds. . . . The reason that they don't open the hatch early and let the crew out is because the crew literally could not walk down the stairs. You can stand, but you cannot walk in a straight line."

"Did you feel crowded with seven astronauts on your last flight?"

Sally nodded and said, "We bumped into each other a lot, spilled each others' food a lot. We came back and recommended that seven was really the maximum crew size without a space module or bigger living space."

After her warm welcome at Stanford, it was on to Hollywood, California, where Sally was to be the recipient of the fourth annual Meridian Award, a citation made each year on the Merv Griffin television show to individuals who had achieved high goals and made new strides in professional and personal endeavors.

The year 1985 brought astronaut Dr. Sally Ride still more public acclaim. In February, in a ceremony held at Disneyland's Magic Kingdom celebrating the thirtieth anniversary of the "Salute to the American Hero" program, she was presented with the American Hero Award for her outstanding achievement and success. In June Sally went to Minneapolis, Minnesota, where Anne Morrow Lindbergh, widow of the famous aviator, Charles A. Lindbergh, presented her with the Lindbergh Eagle Award at a dinner culminating Lindbergh Heritage Week. Even the United States Labor Department paid Sally tribute — by minting two gold medallions in her likeness.

She was soon selected to journey into space for a third time. But then tragedy intervened, altering forever the U.S. space program and all the astronauts in it.

9

The Challenger Disaster

AT 11:30 ON the morning of January 28, 1986, Sally Ride was driving to the space center in Houston, listening to the radio newscaster describing the launch of the tenth shuttle mission of *Challenger*, which was in progress.

The scene at the Kennedy Space Center in Florida was familiar. The ground vibrated as the shuttle's engines roared to life. Great clouds of black and white smoke bellowed from the orbiter as its solid rocket boosters lifted it upward from the launchpad. The spacecraft rolled into position as it headed out over the Atlantic for its final thrust into the heavens.

A voice from Mission Control echoed around the space facility as spectators shaded their eyes from the bright morning light. "Four point three nautical miles, downrange distance three nautical miles." It looked like a picture-perfect launch. But then, invisible to any observers, a flame resembling a small blowtorch crept from the bottom of the right solid rocket booster around the giant orange fuel tank upon which *Challenger* was riding. Mission Control continued, "*Challenger,* go with throttle up."

Commander Francis Scobee replied, "Roger, go with throttle up." These were the last words heard from the astronauts aboard the shuttle *Challenger*. Just 73.23 seconds into the flight, with the craft some ten miles above the launchpad, the shuttle's external fuel tanks exploded. All that could then be seen were many trails of smoke where once a magnificent flying machine had been. There was such shock, surprise, and confusion that no one could be sure what had happened. In fact, the entire vehicle had exploded. Burning spaceship fragments dropped down to Earth for nearly half an hour. All seven astronauts aboard were killed.

Sally was dazed and horrified. The astronauts had been her friends, people whom she respected and cared deeply about. "When I realized how fast it happened, I guess it sort of hit home to me that during a launch that's not what you're thinking about." There was small comfort in knowing that, most likely, the *Challenger*'s crew never realized what had happened.

A wave of shock and grief spread across the nation. President Reagan, scheduled to give his State of the Union address that evening, postponed the event and appeared on television to pay tribute to the lost astronauts. All over the world, people were stunned and filled with sorrow. It was hard to believe that such a tragedy could have happened.

NASA immediately began an investigation of the disaster. Robert Crippen, who had served as commander on both of Sally's flights, was a leader in an early inquiry into the shuttle explosion. He was in charge of the team that was to find and retrieve orbiter fragments which fell to Earth. Thirteen aircraft and more than a dozen ships searched six thousand square miles of ocean looking for remains of the *Challenger*.

Because there was such a public outcry, and, in particular, concern for the fate of the nation's space program in light of this devastating setback, President Reagan set up his own commission to discover why the disaster had occurred and to recommend changes, if any were deemed necessary. The panel was headed by former Secretary of State William Rogers, a New York lawyer, and included only one active astronaut — Dr. Sally Ride. Neil Armstrong, the first person to walk on the moon, was vice-chairman. Other members of the Rogers Commission, as it was called, were scientists, educators, and businessmen. It was President Reagan's hope that such a diverse body would encourage independent views and thus ensure that an objective, unbiased report would be produced. The commission was given 120 days to report its findings. It replaced Captain Crippen's investigation; some thought that a NASA-based inquiry would not be able to maintain the complete objectivity that the task required.

To accommodate the efforts of the commission, for the first few weeks following the disaster NASA astronauts did not discuss the Challenger accident in public. The presidential commission was left to do its work, reviewing the videotapes, photographs, and charts, inspecting shuttle debris, and taking testimony about the prelaunch activities, the lift-off, and the final moments of the flight.

Videotapes revealed a puff of smoke escaping from the lower portion of the right booster rocket as the Challenger was surging aloft. Attention began to focus on the "O-rings," or seals, that were used in the rocket boosters. Sally questioned the safety standards of the shuttle program and zeroed in on the technical procedures of the shuttle and launches. Business people such as Mr. Rogers questioned NASA's decision-making process.

As a result of her knowledgeable probing in the course of the investigation Sally Ride was earning growing respect. She was now called Dr. Ride instead of Sally. During the hearings, she listened with alarming dismay as engineers recounted how they had tried to stop Challenger's launching because they were concerned that the cold weather would affect the sealing capabilities of the O-rings. Realizing that the disaster probably could have been prevented, Sally commented, "It's hard to stop from getting mad."

A grim-faced Sally Ride testifying during the hearings held
in Washington, D.C., to investigate the 1986 Challenger disaster.
The space shuttle program resumed manned missions in
1988 after adopting many of the safety precautions suggested by
the Rogers Commission.

The press revealed that there had been tension between the astronauts and the engineers in the manned space program. Since the disaster, however, it appeared that the astronauts were taking control. Several testified before the commission, and many started to go out and talk to public groups. Interestingly, NASA did not issue special instructions for handling questions about the disaster.

Meanwhile, Sally's book was about to be published. Called *To Space and Back,* it explained in language and pictures suitable for elementary school children what it was she and other astronauts did in space. The shuttle tragedy impelled her to make one important change — the book was to be dedicated not only to her science teacher, Dr. Elizabeth Mommaerts, but also to the memory of her fellow astronauts who were killed when the *Challenger* exploded.

To Space and Back also contains Sally's answer to a question commonly asked by children, and one which seemed especially poignant after the *Challenger* explosion: "Was it scary going into space?"

"All adventures — especially into new territory — are scary," Sally wrote. "And there has always been an element of danger in spaceflight. I wanted to be an astronaut because I thought it would be a challenging opportunity. It was; it was also an experience that I shall never forget."

However, in the wake of the shuttle disaster, she felt compelled to say at one point, "I'm not ready to fly again." Sally had heard too much in the hearings that made her feel insecure about the shuttle's safety. She had discovered that there had been disagreements about whether to launch the shuttle, and even over whether or not it was safe. She had also learned that there may have been flaws in the solid fuel booster rockets.

The Rogers Commission issued its 256-page report in June 1986. Sally thought it was "a significant piece of work. . . because the accident itself was so significant, such an important part of our history." While it contained damning criticism of NASA's safety procedures, engineering faults, and human errors which endangered the astronaut's lives, many people felt that the report hadn't covered all the problems, and that some important facts had been omitted. The report had answered many questions, but also raised others. For example, when it was revealed in many tests that the O-rings were likely to fail in subzero temperatures, why had the astronauts never been informed of the danger? In addition, it was revealed that some engineers had wanted to scrub the launch because of the subzero temperatures at the time of lift-off, citing their tests with the O-rings. Why hadn't their safety concerns been passed along to the launch director who made the final decision for the launch to continue as planned?

The commission was able to determine the exact mechanical cause of the explosion: The right solid rocket booster had sprung a leak which caused the fire; the fire broke through the seal of the rocket booster and led to the explosion. What the commission couldn't fully explain was why it happened.

Even though many such questions remained after the release of the report, it was considered a big first step toward repairing some of the damage to NASA that had occurred as a result of the accident. The report recommended that NASA continue to research the disaster and give further reports to the President. It also gave recommendations for President Reagan to consider, ranging in concern from shuttle design to launching procedures and more.

In August 1986 President Reagan ordered the space agency to get out of the business of commercial satellite launchings — launchings such as the pallet satellite that Sally Ride had worked with on her first flight in space. The private sector, President Reagan asserted, could do a better job at a lower price.

Many of the space shuttle's rocket and safety systems had to be redesigned and tested as a result of the accident. While this was happening, Sally went back to work at NASA headquarters, studying what would be needed for astronaut missions to Mars. The next shuttle flight, it was hoped, would take place in early 1988.

Meanwhile, Sally gave interviews promoting her book, discussed the future of spaceflight, and was inducted into the National Woman's Hall of Fame. In one interview she revealed that she was confident that NASA had done a good job of studying the solid rocket boosters and improving their design and that she would once again consider going into orbit on the shuttle. As it turned out, it would be nearly three years before NASA launched a redesigned shuttle into space. By that time, Sally was looking for new challenges.

10

A Final Contribution to NASA

AFTER SALLY COMPLETED her job on the Rogers Commission, NASA assigned her to be assistant to the agency administrator, a job quite different from that of being an astronaut. For one thing, she was to work behind a desk. The new post was one of the many changes NASA had instituted since the *Challenger* disaster. The Rogers Commission report had recommended that astronauts be moved into management positions where they could make their experience useful to the next group of space scientists. In fact, Sally was one of the first crew members to cross this line. Her responsibilities were to help people at the Long Range and Strategic Planning office to create a plan for NASA's future in space. Sally knew that it was important for NASA to have goals beyond putting a space station into orbit. What exactly were those to be?

For the next eleven months Sally led a team of researchers in conducting a close study of the entire space agency. It was Sally's aim to produce a report that would give safe guidelines for the future of space exploration. In August 1987 she issued "Leadership and America's Future in Space," a sixty-three-page analysis giving four options for the future of the space program. Sally did not, as was customary, hold a news conference to discuss the report. Rather, she felt, the work would speak for itself. It did. Sally's leadership of the research team was quickly recognized, and the document quickly became known as the Ride Report.

The space program needed new direction, the Ride Report explained. It had been nineteen months since the *Challenger* disaster, and it was time for the space agency to move forward. Sally felt that without an eye on the future, the space program would flounder in the present.

First, the report pointed out that the U.S. space program had lost its leadership in space to the Soviet Union. Even though NASA had pioneered the exploration of the planet Mars, no new spacecraft had been sent there since 1976. The Soviets, on the other hand, had developed an aggressive program to explore the surface of Mars, and were even starting to use robots in their exploration of the red planet.

Although NASA had been concentrating on the commercial use of space, no space stations had been sent into orbit. Indeed, NASA hadn't sent astronauts to live in space since 1974, when Skylab was used. Meanwhile, the Soviets had put eight space stations in orbit since the mid-1970s. In pointing out these weaknesses, the Ride Report demonstrated that there were other important programs that NASA needed to improve besides the space shuttle.

The first option given in the Ride Report was named "Mission to Planet Earth." It called for a program of satellites operating at low orbits to study the Earth to analyze the changes that had taken place in the planet's environment as a result of man's activity. Sally, as a scientist, had come to realize that the Earth was a much more fragile planet than most people thought. For example, changes on the surface that occurred when cities replaced forests, or those which occurred in the interior of the planet from underground testing of nuclear weapons, could drastically alter the environment, much to the detriment of the life-forms — humans included — inhabiting the planet. Pollution was yet another problem. Because these were worldwide problems not restricted to one nation, Sally recommended the formation of an international project to study the effect of mankind's activity on the Earth's environment.

Exploration of the solar system was the second option recommended in the Ride Report. Sally wanted the United States to regain its leadership in space and thought that sending an unmanned spacecraft to explore an "outer" planet, such as

Saturn, and an "inner" planet, such as Mars, would provide the means. The idea of exploring Mars had always seemed to stimulate public imagination. Sally understood that public and government support for the space program could not rely solely on the fact that scientific research from space was used to create valuable new discoveries on Earth.

The third option proposed in the Ride Report was called "Outposts to the Moon" and recommended that some permanent or semipermanent station be established on the moon's surface by the year 2005. By 2010, the report said, it might even be possible for people to live on the moon for months at a time.

"Humans to Mars" was the last option offered. The report discussed the chances of Americans landing on and exploring the planet Mars by the beginning of the twenty-first century. Here, too, building an outpost of some sort was one of the possibilities under consideration.

It is important to note that Sally was not trying to start a race to Mars. Rather, she stressed, it was important to explore those planets close to Earth and to gain experience living in space. Overall, she emphasized a cautious, one-step-at-a-time approach, placing primary importance on the safety of astronauts, not on beating the Soviet Union in a space race.

Sally believed that the Ride Report had sent out a strong message. She felt that President Reagan and the public had a choice: either lessen the demands on the space agency or make more money available so that the work could be done more safely.

11

Sally Ride: A Modern Heroine

ON MAY 26, 1987, Sally Ride celebrated her thirty-sixth birthday. On that same day, NASA announced that she would be leaving the astronaut corps — she had accepted a new job at Stanford University, her old alma mater, to work at the university's Center for International Security and Arms Control, which had a program for midcareer scientists to train in national security matters.

"I've always wanted to go back to a university setting," Sally said. "I've spent many happy years at Stanford, as a student and a graduate. I just got the right offer." Still, Sally never revealed exactly why she was changing jobs. As usual, she gave the press few details. "I am confident of the future of our nation's space program," she told them, "and will always remember my friends at NASA. The various projects I have participated in have provided unique challenges and have allowed me to grow as a scientist and a person. It is in that same spirit of challenge that I have accepted a position at Stanford University."

There was much speculation as to why Dr. Sally Ride was leaving. Some thought the question of when she would be able to resume flying might be responsible for Sally's sudden change in her future career plans. Although Sally was still listed as being on active flight status, she felt that she had become firmly entrenched in a management position and that in reality it could be years before she would be selected for another shuttle flight. Furthermore, the possibility of Sally's flying in space again was even more unlikely when she realized that NASA wasn't planning to launch satellites which would require using the robotic arm — Sally's particular area of expertise. In addition, Sally's divorce from her astronaut husband led some to speculate that she was in need of a change of scenery. Whatever her reasons, Sally left the public eye in 1987 to return to life as a private citizen and scientist.

Her departure from the astronaut corps left eleven women astronauts ready to fly out of a total of eighty-three. Sally wasn't the first to leave NASA since the *Challenger* disaster — ten other astronauts had departed as well.

Summing up Sally's contribution to NASA, Agency Administrator James Fletcher said, "The nation owes her a debt of gratitude. Her flight as the first woman in space firmly established an equal role for women in the space exploration program. . . . Today, the assignment of women to shuttle crews is a routine matter based on ability and need and is no longer a cause for notice. . . . The country is fortunate that her energy, intelligence, and good sense will continue to be focused on matters of vital public interest."

Someone once asked her, "What will you remember most about your flight in space?" "The thing that I'll remember most about the flight," Sally said with a smile, "is that it was fun. In fact, I'm sure that it was the most fun that I will ever have in my life."

Afterword

America in Space:
A Brief History

THE SPACE AGE began on October 4, 1957, when the Soviet Union launched *Sputnik 1,* the first earth-orbiting satellite. Despite this remarkable scientific achievement, the American public was profoundly shocked and dismayed — its self- image and faith in America's superior technology had been shattered by the fact that the Soviets had "gotten there first."

Americans, accustomed to being first in most everything, waited anxiously for the nation's scientists to demonstrate that U.S. space technology could equal and surpass that of the Soviet Union. But the subsequent months were bitterly disappointing for the nascent American space program. First, in November, the Soviet Union followed up the successful *Sputnik* mission by lofting a second satellite, this time with a canine passenger! The first U.S. attempt to launch a satellite was quickly scuttled when, just two seconds after ignition on December 6, 1957, the needle-shaped *Vanguard* rocket exploded in a ball of flames. This effort was followed by a second failure. Finally, on January 31, 1958, the U.S. Army launched *Explorer 1*. America was, at last, in space.

Explorer 1 was a pencil-shaped satellite that weighed about thirty pounds and was equipped with radio transmitters and scientific instruments that measured radiation, temperature, and the frequency of tiny meteor collisions with the spacecraft. During its time in space, *Explorer 1* made two important discoveries. First, it confirmed the existence of a huge belt of radioactive particles trapped by the Earth's magnetic field, now named the Van Allen Radiation Belt. Second, it dispelled the fear that minuscule meteors traveling at high speeds near Earth could puncture an orbiting spacecraft or satellite. In fact, there were only seven small meteor "hits" in the satellite's first month in orbit.

The Soviet Union also beat the United States in putting the first human into space. On April 12, 1961, cosmonaut Yuri Gagarin orbited the Earth once aboard *Vostok I* in a flight that lasted one hour and forty-eight minutes. Less than a month later, on May 5, Alan B. Shepard, Jr., became the first American to venture into space. Enclosed inside *Mercury 4*, a small capsule perched atop a

Redstone missile, Shepard's historic, though suborbital, flight took him to a peak altitude of 115 miles. Then he and the capsule landed safely in the Atlantic Ocean off the coast of Florida. Shepard's only complaint about the fifteen- minute ride was that "It didn't last long enough."

America's first orbital flight came on February 20, 1962, after weeks of frustrating delays. John H. Glenn, Jr., a freckled Marine Corp test pilot, was launched inside the cramped *Mercury 6* capsule, named *Friendship 7*. He circled the Earth three times, reentered the Earth's atmosphere, and splashed down to national rejoicing.

Over the next fifteen months three more Mercury flights vaulted astronauts into orbit in preparation for what had become the nation's number one space objective: to land an American on the moon by the end of the decade. President John F. Kennedy had elucidated this goal upon taking office in 1960, and his spirit had infused the National Aeronautics and Space Administration (NASA) and the public at large with a compelling sense of purpose.

While much of America's space efforts were aimed at achieving a manned lunar landing, unmanned space probes laden with sophisticated instruments scouted the solar system for data that would, it was hoped, help answer a number of broad scientific and philosophical questions. How far can humans travel? What kinds of things are out there in space? Can other planets support life as we know it? Are there other kinds of life in the universe?

The Ranger program concentrated on moon exploration. In July 1964 *Ranger 7* took and transmitted back to Earth the first photographs of the surface of the moon. More than seventeen thousand photographs were amassed, all of which were analyzed and used in preparation for the manned lunar landings which followed. The probes were sent on what amounted to suicide missions, taking pictures and transmitting data to Earth until the moment they crashed — as programmed — into the lunar surface.

The Mariner program consisted of ten unmanned space probes which were to "fly-by" or orbit Mercury, Venus, and Mars, the planets closest to our own. In August 1962 *Mariner 2,* passed within 21,648 miles of Venus and sent back data indicating that the planet's surface temperature was 800 degrees Fahrenheit, too hot for any known plant or animal. *Mariner 2* continued to send information back to Earth until January 1963, when the ship had journeyed 54 million miles away from Earth and radio contact was lost. *Mariner 4* was sent toward Mars, and in July 1965 provided the first photographs of the Martian surface.

Gemini, an extension of the Mariner program, featured a series of manned flights designed to tackle the complex problems of prolonged space flight — journeys with a duration of two weeks or longer. In June 1965, while *Gemini 4* commander James McDivitt looked on from inside the spacecraft, fellow crew member Edward White became the first American to walk in space. Connected to a twenty-seven-foot tether, White floated in space for twenty minutes as he and his craft orbited Earth. McDivitt's efforts were later bettered by *Gemini 12*'s Edwin "Buzz" Aldrin, Jr., who worked outside his spacecraft for five-and-a-half hours without tiring. *Gemini 5*'s eight-day flight proved that people could survive in space long enough for a round-trip to the moon. *Gemini 6* and *Gemini 7* achieved the first successful rendezvous of two spacecraft. *Gemini 11* demonstrated command of a variety of sophisticated docking maneuvers. Clearly, as the decade progressed, the American manned space program was taking bold steps toward realizing the dream of a moon landing.

Meanwhile, unmanned probes continued to explore the moon's surface. In June 1966 *Surveyor 1* landed on the moon — offering the first proof that the lunar surface was hard enough to support the weight of a spacecraft — analyzed its soil, and sent back ten thousand photographs that were used to select sites for the planned moon landing. By January 1968 seven more Surveyor spacecraft had successfully landed on the moon.

The outer reaches of the solar system were also within reach of U.S. interplanetary probes. The Pioneer program, started in the early 1960s, was the first series of unmanned spacecraft that made important discoveries about Saturn and Jupiter. In 1986, fourteen years after it was launched from Earth, *Pioneer 10* became the first artificial object to leave the solar system.

Still, the moon remained most everyone's main curiosity, as it has since the beginning of time. The Apollo program was started to land a man on the moon and safely return him to Earth. It took eleven well planned and executed missions, and one tragic accident that cost three astronauts their lives, to accomplish the task. On July 20, 1969, astronauts Neil A. Armstrong and Buzz Aldrin became the first humans to set foot on the moon. After successfully answering President Kennedy's challenge to land a man on the moon by the end of the decade, NASA launched another six Apollo missions before the program ended in 1975.

In 1973 NASA launched Skylab, America's first manned orbiting space station, part of a program designed to show that man could safely live and work in space for prolonged periods. Skylab, at over 118 feet in length, was as big as a small house. Weighing in at more than seventy-seven tons, it was made up of three sections: the orbital workshop, an air-lock module, and a multiple docking adapter. There was also a telescope mount. The interior space was divided into two floors, the lower one for crew quarters, medical experiments, exercise equipment, and air lock, and a spacious, domed upper floor for use as a general storage and activity area. Three three-man crews were ferried to the station and stayed in orbit for twenty-eight days, fifty-nine days, and eighty-four days, observing solar activity and evaluating systems designed to gather data about Earth's resources and environmental conditions. Eventually, and after the three missions had been completed, Skylab began to descend from its orbit. In July 1979 most of it disintegrated upon reentering Earth's atmosphere; the parts that did not burn up fell on Australia and into the Indian Ocean without causing injury or damage.

NASA's unmanned space program took another leap forward in 1975 with the launching of *Viking 1* and *Viking 2*. Their target was Mars — the "red planet" of age-old fascination, thought by scientists and laypeople alike to be the best candidate for life in the solar system in addition to the Earth. Both probes reached their goal and sent back stunning color pictures — the first photographs ever taken from the surface of another planet. Mars, it turned out, had a rocky surface, a pink sky — scientists had thought it would be deep, dark blue — and land formations that seemed to indicate the previous existence of water. Temperatures around the *Viking* landing sites ranged from minus 123 degrees Fahrenheit at night to minus 20 degrees Fahrenheit during the daytime. Winds blew at six to eight miles per hour, with gusts of up to thirty to forty miles per hour — much gentler than NASA scientists had expected. The question of whether there is or ever has been "life on Mars" went unanswered.

Viking's inability to detect signs of life may mean that the planet's atmosphere is indeed hostile to life as we know it — but it might also mean that the experiments were designed incorrectly.

Whereas the *Viking* project saw NASA trying to find clues about the evolution of life, Project *Voyager* has as its focus the origin and evolution of the solar system itself. Launched in 1977 and planned as a series of outer planet "fly-bys," *Voyager 1* and *Voyager 2* have transmitted back to Earth information and images that have expanded and revolutionized our knowledge of the planets. At Jupiter in 1979 *Voyager* discovered what may be an underground ocean and the first active volcanoes outside Earth. At Saturn in 1981 *Voyager* found thousands of rings, not just the few that had been expected. Its visit to Uranus in 1986 was the space program's first and revealed a stream of wonders, including 15 moons and pitch-black rings. Using "swing-by trajectories" that take advantage of the outer planets' gravitational forces, the two *Voyager* craft have been propelled further into the solar system and are on their way to encounters with Neptune, Pluto, and beyond. All in all the probes have sent back more than four trillion bits of information, the equivalent of about 100,000 encyclopedia volumes! Unless some vital system failure occurs, the Project *Voyager* will be returning data to Earth well into the twenty-first century.

On April 12, 1981, NASA launched the space shuttle *Columbia,* inaugurating the first reusable space transportation system. Other shuttles — the *Challenger, Discovery,* and *Atlantis* — made their debuts in the next few years, forming a fleet which went on to conduct twenty- four successful missions through the end of 1985. Most of the shuttle cargoes consisted of communication, defense, and weather satellites and self-contained experiments. The spacecraft can also be used as an Earth-orbiting space station for about thirty days, though it has not yet been used in this fashion.

All went well with the shuttle program, at least on the surface, until January 28, 1986, when the *Challenger* exploded just seventy-three seconds after lift-off, killing all seven astronauts aboard. The accident was the world's worst spaceflight disaster and sent America's space program into a tailspin. After an official investigation into the accident, the cause of the disaster was found to be a defective seal in the solid rocket booster. It was also concluded that the accident had happened because of intense pressure for increasing the number of launches in a given period, poor internal management and communication at NASA, and, perhaps most disturbing, the fact that the warnings of several flight engineers concerned about the below-freezing temperature on the morning of the launch had been ignored.

By the end of 1986, while many of the shuttle's systems were being redesigned, several officials from NASA and Morton Thiokol (the manufacturer of the solid rocket booster) had resigned. During this time the United States, attempting to restart its floundering space program, launched four types of large disposable rockets, all of which had been grounded by failures earlier in the year. Finally, after months of delays, the shuttle *Discovery* blasted off on September 29, 1988. After a thirty-two month absence, the United States was back in space. This near-perfect mission was followed by the launching of the shuttle *Atlantis* on December 2, 1988. At least for the time being, the space program was back on track. Still, lively debate continues to surround NASA's future. Should there be a manned mission to Mars? Should the United States construct a permanent Earth-orbiting space station or a lunar colony? Should manned missions be terminated altogether in favor of computer-operated unmanned probes? Should space research and exploration be funded by the government, private enterprise, or some combination? Should the United States and the Soviet Union share their space technology? These are just some of the complex issues that NASA administrators and other officials will have to grapple with in the years to come. The decisions that are made today will affect America's space program for many years to come, but one thing is for certain — our fascination with the moon, stars, and planets is ancient, and space exploration is here to stay.

Appendix

U.S. Space Camp:
Where to Learn the Right Stuff

DID YOU KNOW that there is a camp where ordinary people can learn what it takes to become an astronaut? U.S. Space Camp, located in Huntsville, Alabama, is just such a place — and you don't have to have a straight "A" average, be a brain in science, or a super athlete to be admitted to the program. All you need is the desire to learn and a willingness to do a job the best you can.

U.S. Space Camp was the brainchild of the late Wernher von Braun, the German scientist and rocket pioneer who played a large and influential role in the development of America's space program. As he once said, "There are tennis camps, basketball camps, and cheerleader camps. Why not a space camp?"

Fittingly, U.S. Space Camp is located adjacent to the National Aeronautics and Space Administration's Marshall Space Flight Center, where real-life astronauts are trained. Also close at hand is the U.S. Army's Redstone Arsenal, where rocket engines are developed and tested. It is part of the 450-acre Space and Rocket Center, America's largest showcase of space technology. The center is widely noted for its "learn by doing" exhibits related to astronaut training and rocket technology.

The camp opened its doors in 1982. Since then interest in its programs has, well, skyrocketed — from 757 campers in its first year to 19,000 in 1988. Says Camp Director Edward Buckbee, a former NASA employee and colleague of von Braun's: "The [kids who attend this camp] will be the space-flying generation. They're excited about making a contribution to the space program. And we expect that some of them will man [NASA's] new space station, will help design and establish a lunar colony, and will be aboard the manned mission to Mars."

Just what goes on at Space Camp? Since the programs are designed for different age groups, that depends on what grade you're in.

Space Camp is open to boys and girls in grades four through six. Over the course of five days, "trainees" build and launch model rockets, perform science experiments, watch space movies, attend lectures, ride in jet simulators, and romp

112

in a simulated lunar environment called Rocket Park.

Space Academy I is open to boys and girls in grades seven through nine. This is also a five-day program, during which the trainees work in a shuttle cockpit simulator, a full-scale Spacelab module, and a reproduction of Houston's Mission Control. Specific mission assignments are given, offering a chance to perform both onboard activities and ground support functions.

Space Academy II is open to young men and women in grades ten through twelve who have a serious interest in science. These students spend ten days at Space Camp scuba diving in the *Neutral Buoyancy Tank,* performing complex experiments, and more. This advanced program is patterned from NASA crew training manuals and features a series of two-hour missions in preparation for a 24-hour simulated space shuttle mission.

Among the highlights for everyone is the *Microgravity Training Chair,* which was used during the Apollo program to train astronauts for walking on the moon. The chair hangs on springs; when properly weighted and balanced, the springs offset five-sixths of the trainee's weight, giving the sensation of walking in the moon's lighter gravitational pull. Another is the *Five Degrees of Freedom Training Simulator.* Used in the early Gemini program, the "5DF" is a spacewalk simulator. Spindles attached to the seat provide freedom to move on two axes. The

113

apparatus floats on air bearings a fraction of an inch above the floor, giving trainees freedom to move from front-to-back, side-to-side, or to spin on the third axis.

In addition, everyone gets to eat space food, wear (and be photographed in) flight suits, go through push-up drills at 0700 hours (don't forget, this is astronaut training!), and receive diplomas and a nifty set of astronaut wings. NASA also sends real astronauts over to meet and talk with the trainees.

U.S. Space Camp also has a dormitory called the Habitat whose design was influenced by that of NASA's planned space station. Longer than a football field, sheathed in a silver-tone frame, and complete with portholes, ladders, latches, sleep stations built into horizontal tubes, computer workstations, and more, the Habitat can accommodate 444 trainees and counselors. Its space-age look symbolizes the overall feeling of U.S. Space Camp — the countdown to the twenty-first century has already started!

DATA BANK
* Fifteen percent of the trainees are girls.
* There is a second U.S. Space Camp in Titusville, Florida, near the Kennedy Space Center.
* Tuition rates at both sites range from $425 to $800, depending on the time of year (the camp operates in spring, summer, and fall) and program one attends. Included are housing, meals, all materials required for the programs, and tours

114

of the space centers — Marshall in Alabama and Kennedy in Florida.

* Scholarships are available and are based on scholastic achievement, ethnic background, and financial need. Candidates are required to write an essay on a space-related topic and must be sponsored by a teacher who verifies the applicant's interest in space exploration and science. * Completion of each track (Aerospace, Engineering, and Technology) earns one hour credit of freshman-level science from the University of Alabama in Huntsville.

* A program for adults is also available.

* U.S. Space Camp is a nonprofit educational venture.

For more information, write to:
U.S. Space Camp
The Space & Rocket Center
Tranquility Base
Huntsville, Alabama 35807
Or call: (205) 837-3400

GLOSSARY

aerodynamic — relating to the force of air in motion; the science of *aerodynamics* examines what happens when air or any other gas is in motion

astrophysics — the science which studies the physical properties and phenomena of the stars, planets, and other objects in space

calibrating — dividing or marking levels of degree or quantity on a measuring instrument, such as a thermometer

"capcom" — abbreviation for capsule communicator. During space missions, one person on Earth — the capsule communicator — is assigned the task of being the communications link with the crew.

centrifuge — a machine that turns at a high speed to create *centrifugal force,* which causes substances of different densities to separate. A milk separator, which separates cream from milk, is a good example. Centrifuges also produce different gravitational effects, and so are used as part of space training to duplicate the conditions that astronauts will experience in space.

free-electron laser — a special kind of laser that makes it possible to identify objects in space that were previously undetectable

g-force — a force similar to gravity that is produced by the acceleration of a rocket

greenhouse effect — the warming of the Earth's atmosphere due to the accumulation of certain gases, such as carbon dioxide. Many scientists say that the greenhouse effect could lead to major, and possibly damaging, changes in the Earth's environment in coming years.

mass spectrometer — a device that uses light to detect and identify the different kinds of particles present in a substance; also called a *mass spectrograph*

metabolism — the physical and chemical processes continuously occurring within living organisms and cells by which food is turned into energy and waste materials

microbe — a very small (microscopic) living thing, plant or animal; also called a *microorganism* or *germ*; also used to refer to disease-producing bacteria

mission control — the central communications station for a space mission, separate from the launching station. Astronauts receive orders from mission control and report in to mission control about the progress of the mission.

physiology (physiologically) — the branch of biology dealing with the functions of living things or their parts (organs, tissues, cells)

propellant — a fuel substance used to propel a rocket

simulate/ simulated — to look or act like, to have the appearance or characteristics of. Training programs for everything from space travel to automobile driving use imitations of the real machines to prepare for the actual experience.

space adaptation syndrome — the scientific term given to *space sickness*, an illness experienced by many astronauts because of changes in atmosphere and motions in space

Other books you might enjoy reading

1. Behrens, June. *Sally Ride, Astronaut.* Childrens Press, 1984.

2. Blacknall, Carolyn. *Sally Ride.* Dillon Press, Inc., 1984.

3. Cross, Wilbur and Susanna. *Space Shuttle.* Childrens Press, 1985

4. Fox, Mary Virginia. *Women Astronauts Aboard the Shuttle.* Julian Messner, 1987.

5. Kerrod, Robin. *Space Shuttle.* Gallery Books, W.H. Smith Publishers, 1984.

6. Machnight, Nigel. *Shuttle.* Macknight International, 1985.

7. McGowen, Tom. *Album of Space Flight.* Rand McNally & Co., 1983.

8. O'Conner, Karen. *Sally Ride and the New Astronauts: Scientists in Space.* Franklin Watts, 1983.

9. Ride, Sally. *To Space and Back.* Lothrop, Lee & Shepard Books, 1986.

10. Wilson, Andrew. *The Shuttle Story.* Hamlyn Publishing, 1986.

ABOUT THE AUTHORS

Jane Hurwitz earned her M.A. degree at Kansas University. In addition to her writing for young readers, she works as an economist. She lives in Connecticut.

Sue Hurwitz has a M.A. in education and taught school for many years. She has published dozens of stories for children.